German Armored Warplanes of WWI

A Centennial Perspective on Great War Airplanes

Jack Herris

Great War Aviation Centennial Series #4

This book is dedicated to my aunt and uncle, Dorothea and Herb Evans, who have been a great inspiration to me.

Acknowledgements
My sincere thanks to Reinhard Zankl for his insightful information and comments about production orders and serial numbers, to Colin Owers, Aaron Weaver, Paolo Varriale, and Boris Ciglic for photographs, and Greg VanWyngarden for photographs and his very helpful suggestions and material. Thanks also to Jerry Boucher for his original painting done for this book, Bob Pearson for his color profiles, Aaron Weaver for his digital photo editing, Martin Digmayer for his scale drawings, Jim Miller for his colorized Albatros J.I photo, and the Deutsches Technikmuseum in Berlin for photographs. Any errors are my responsibility.

Cover painting by Jerry Boucher. Please see his website: **www.the-vaw.com**

Color aircraft profiles © Bob Pearson. Purchase his CD of WWI aircraft profiles for $50 US/Canadian, 40 €, or £30, airmail postage included, via Paypal to Bob at: **bpearson@kaien.net**

For our aviation books in print and electronic format, please see our website at: **www.aeronautbooks.com**. You may contacted me at **jherris@verizon.net**.

For our aviation books in print and electronic format, please see our website at: **www.aeronautbooks.com**. I am looking for photographs of the less well-known German aircraft of WWI. For questions or to help with photographs you may contact me at **jherris@verizon.net**.

ISBN: **978-1-935881-11-7**

Design and layout: Jack Herris
Cover design: Jack Herris
Digital photo editing: Jack Herris & Aaron Weaver

Aeronaut Books

www.aeronautbooks.com

Table of Contents

Introduction

Above: Crewmen pose in front of Junkers J.I J.115/17. The Junkers J.I was the world's first all-metal production aircraft and the toughest J-type; the J.I was known to return home with 200mm (8") holes in the wings!

The German air service experienced a difficult year of transition in 1916. During the Battle of Verdun (February 21 to December 18, 1916) the French achieved air superiority over the German air service by introduction of technically superior fighters (the Nieuport 11), better tactics (combining fighters into squadrons to concentrate their power), and a superior operational concept (concentrating warplanes for offensive missions while Germany initially relied on ineffective barrier patrols). The British then launched the Battle of the Somme (June 24 to November 26 1916) before the German High Command was able to adjust their aviation forces to address their shortcomings.

By July 1916 new German Halberstadt biplane fighters started arriving at the front, followed by new Albatros fighters in August, along with organization of the first German *Jadgstaffeln*, or fighter squadrons, which became effective in September. By October the *Luftstreitkräfte* (German Air Service) was completely reorganized, the *Jastas* were regaining air superiority, and the crisis was past.

Birth of the German Infantry Aircraft

Part of the overhaul of the new air service was a comprehensive review of its performance in the great battles of 1916. Based on their experience of the Battle of the Somme, in which communications with front-line troops were often thrown into chaos by the confusion of battle, telephone lines severed by artillery fire, and runners who were often unable to complete their missions, all before development of effective radios, the German High Command concluded that low-flying reconnaissance aircraft could quickly provide critical information about the battle. This information, essential to rear-area staffs directing the battle, included the location of German and enemy troops, enemy strong-points, and movements behind enemy lines that signaled enemy intentions.

At first, standard C-type aircraft were tasked to cooperate with the infantry, and initial use of *Infanterieflieger* (infantry flyers) in the Spring of 1917 gave good results. Crews fastened colored pennants to the standard C-type aircraft to enable German troops to recognize them so they would not fire on them. To show their position, the German troops laid out white identification panels, fired signal flares, and communicated with signal lamps. Soon the *Infantrieflugzeug* would return to base and the crew could deliver exact positions and other battle information to the division command via photographs, sketches, and oral reports. Unsurprisingly, Allied defenses improved and losses of infantry fliers in their unarmored aircraft rose sharply by mid-1917.

The *Luftstreitkräfte* had anticipated this situation, and in November 1916, *Idflieg* (*Inspektion der Fliegertruppen* — Inspectorate of Aviation Troops) issued a specification calling for an armored infantry support aircraft designed so the pilot and gunner, engine, and fuel were completely enclosed by 5mm steel armor for protection against rifle-caliber ground fire, which included machine guns. *Idflieg* did not consider performance of these armored aircraft to be important for self-defense, so they were not required to climb over 1,500 meters. Similarly, *Idflieg* did not require high speed because, in theory, their armor

Frontbestand Class J Armored Infantry Aircraft Frontline Inventory

		1914			1915						1916						1917						1918			
Manufacturer and Type		31 Aug	31 Oct	31 Dec	28 Feb	30 Apr	30 Jun	31 Aug	31 Oct	31 Dec	28 Feb	30 Apr	30 Jun	31 Aug	31 Oct	31 Dec	28 Feb	30 Apr	30 Jun	31 Aug	31 Oct	31 Dec	28 Feb	30 Apr	30 Jun	31 Aug
AEG	J.I																		4	1	11	35	55	66	43	28
	J.II																							15	65	63
	J.																		3							
Albatros	J.I																					37	42	33	51	16
	J.II																									19
Junkers	J.I																			1	1	4	16	25	25	60
Total:																			7	2	12	76	113	139	184	186

would protect these aircraft. In practice the aircrew flying these armored warplanes did not fully agree with *Idflieg* and generally wanted more speed, climb, and maneuverability.

The letter 'J' was chosen to represent the abbreviation of *Infantrie Flugzeug* (infantry aircraft) to avoid confusion with the Roman numeral 'I' because the letters 'I' and 'J' are often interchangeable in German writing. The armored warplanes thus became J-types.

The armament specified was a flexible machine gun for the observer and, initially, a fixed machine gun for the pilot. However, *Idflieg* considered armored aircraft too slow and cumbersome to successfully attack enemy aircraft with the forward-firing gun. Furthermore, to attack ground targets the pilot would have to go into a shallow dive, which *Idflieg* considered too dangerous for such ponderous airplanes at low altitude. Therefore the requirement for a forward-firing gun was eliminated before any J-types were built, thus saving weight, an important consideration due to the heavy armor.

After circulating the J-type specification, *Idflieg* chose AEG, Albatros, and Junkers to build prototypes. Each company had certain advantages and all three responded with designs that were placed in production. The AEG company's advantages included their familiarity with robust, fabric-covered metal structures that were relatively easy to manufacture, plus a substantial production capacity, and their J-type aircraft reached the front first. The Junkers' advantage was tough, all-metal construction using Duralumin, and the Junkers J.I, next to reach the front, went on to a very successful operational career as the best of the J-types. In addition, the Junkers J.I earned a place in history as the world's first all-metal production aircraft. The Albatros company's advantages included their huge production capacity and familiarity with wood construction, a material more readily available in blockaded Germany than duralumin. Due to their wooden construction the Albatros J-types were not as resistant to ground-fire as the other designs, so were less popular with the aircrews. They were last to reach the front, and served in the smallest numbers.

Evolution of German Armored Warplanes

According to *Idflieg* the primary task of J-type aircraft was to provide rear-area staffs with information about troop positions and the flow of battle. To fulfill this role, low-level photography equipment and a wireless transmitter were standard equipment in J-types.

In contrast, *Schlachtflugzeuge* (battle aircraft, primarily CL-types) were designed to attack enemy troops and associated ground targets. To fly on such patrols was called *Iffeln* (*Infantriefliegen*, pronounced '*iffelif*') in German airmen's slang, regardless of which aircraft type they flew. The British called this work 'infantry contact patrols'; today it is more generally called close air support and ground attack.

As the *Infanterieflieger* gained combat experience with J-type aircraft, they started to use J-types for ground attack in addition to their designed infantry cooperation role, resulting in the call for more effective offensive armament. This led first to installation of fixed, downward-firing machine guns for strafing in the AEG and Albatros J-types, and later to flexible 20mm Becker cannon in those types for tank-busting as Allied tanks became ever more of a problem for the German army. The Junkers J.I , the most robust aircraft in the infantry cooperation role, was too heavy and under-powered to carry additional armament and relied on the observer's flexible machine gun throughout its career. With its greater offensive armament the AEG J.II was the most effective J-type for ground-attack.

Several other armored warplane types were designed and built too late for combat. One was an armored fighter to intercept Allied ground-attack at low altitude. Another was the new S-type, an armored type optimized for anti-tank missions.

J-Type Specifications					
	AEG J.I	**AEG J.II**	**Junkers J.I**	**Albatros J.I**	**Albatros J.II**
Engine	200 hp Benz Bz.IV	200 hp Benz Bz.IV	200 hp Benz Bz.IV	200 hp Benz Bz.IV	220 hp Benz Bz.IVa
Span, Upper	13.00m (42.65 ft.)	13.00m (42.65 ft.)	15.0m (49.2 ft.)	14.14m (46.4 ft.)	14.1m (46.25 ft.)
Span, Lower	12.48m (40.95 ft.)	12.48m (40.95 ft.)	—	13.45m (44.13 ft.)	13.45m (44.13 ft.)
Chord, Upper	1.65m/1.85m (5.41 ft./6.07 ft.)	1.85m (6.07 ft.)	2.25–2.5m (7.38–8.20 ft.)	1.70m (5.58 ft.)	1.70m (5.58 ft.)
Chord, Lower	1.65m/1.85m (5.41 ft./6.07 ft.)	1.85m (6.07 ft.)	1.08–1.5m (3.54–4.92 ft.)	—	—
Gap	1.95m (6.40 ft.)	1.95m (6.40 ft.)	2.00m (6.56ft.)	—	—
Stagger	0.12m (0.39 ft.)	0.07m (0.23 ft.)	—	—	—
Length	7.20m (23.62 ft.)	7.86m (25.79 ft.)	9.06m (29.72 ft.)	8.83m (29.0 ft.)	8.44m (27.75 ft.)
Height	3.3m (10.83 ft.)	3.3m (10.83 ft.)	3.47m (11.38 ft.)		
Track	2.19m (7.19 ft.)	2.15m (7.05 ft.)	—	2.0m (6.56 ft.)	2.0m (6.56 ft.)
Wing Sweepback	0	0	—	2°	1.5°
Wing Area	39.0 m^2/43.6 m^2 (420 $ft.^2$/469 $ft.^2$)	43.6 m^2 (469.3 $ft.^2$)	50.84 m^2 (547.2 $ft.^2$)	42.82 m^2 (462.4 $ft.^2$)	43.2 m^2 (466.6 $ft.^2$)
Empty Weight	1,456 kg (3,210 lb.)	1,480 kg (3,263 lb.)	1,766 kg (3,893 lb.)	1,398 kg. (3,075 lb.)	1,515 kg. (3,340 lb.)
Loaded Weight	1,876 kg (4,136 lb.)	1,900 kg (4,189 lb.)	2,176 kg (4,797 lb.)	1,808 kg. (3,978 lb.)	1,927 kg. (4,249 lb.)
Maximum Speed	140 km/h (87.5 mph)	140 km/h (87.5 mph)	155 km/h (96.3 mph)	140 km/h (87.5 mph)	140 km/h (87.5 mph)
Climb to 1,000m	6.1 minutes	6 minutes	12.5 minutes	11.2 minutes	11.2 minutes
Climb to 2,000m	15.9 minutes	14 minutes	33.8 minutes	—	—
Climb to 3,000m	31.6 minutes	30 minutes	77.0 minutes	—	—
Flight Duration	—	—	—	2.5 hours	2.5 hours
Armament	1 flexible machine gun, some aircraft 2 fixed machine guns	1 flexible machine gun, 2 fixed machine guns	1 flexible machine gun	1 flexible machine gun, some aircraft 2 fixed machine guns	1 flexible machine gun, 2 fixed machine guns
Optional	—	20mm Becker Cannon	—	20mm Becker Cannon	20mm Becker Cannon
Note: AEG J.I specs are shown as early/late; later production AEG J.I aircraft had 1.85m chord and wing area of 43.6 m^2. AEG J.I dihedral = 2°					

Finally, an armored variant of the AEG G.IV, the AEG G.IVk, was also designed for anti-tank duties. Both the S-types and AEG G.IVk were armed with flexible 20mm Becker cannon to destroy tanks after successful anti-tank operations by cannon-armed AEG and Albatros J-types. WWI tanks were armored against rifle-caliber fire but 20mm cannon easily penetrated their armor, especially from above. These cannon-armed aircraft presaged the modern anti-tank helicopter.

Miscellaneous Armored Aircraft Specifications				
	AEG DJ.I Triplane	**AEG DJ.I Biplane**	**AEG G.IVk**	**Ago S.I**
Engine	195 hp Benz Bz.IIIb V-8	195 hp Benz Bz.IIIb V-8	Two 260 hp Mercedes D.IVa	260 hp Basse und Selve BuS.IV
Span Upper	11.20 m (36.75 ft.)	10.00 m (32.81 ft.)	18.00 m (59.06 ft.)	—
Span Middle	10.60 m (34.78 ft.)	—	—	—
Span Lower	10.00 m (32.81 ft.)	9.90 m (32.48 ft.)	17.50 m (57.41 ft.)	—
Chord Upper	1.475 m (4.84 ft.)	1.80 m (5.91 ft.)	2.40 m (7.87 ft.)	—
Chord Middle	0.825 m (2.71 ft.)	—	—	—
Chord Lower	0.825 m (2.1 ft.)	1.60 m (5.25 ft.)	2.40 m (7.87 ft.)	—
Gap	1.00 m (3.28 ft.)	2.00 m (6.56 ft.)	2.60 m (8.53 ft.)	—
Wing Area	31.3 m^2 (336.8 $ft.^2$)	30.5 m^2 (328.2 $ft.^2$)	65 m^2 (699.4 $ft.^2$)	—
Length	—	6.90 m (22.64 ft.)	9.00 m (29.53 ft.)	—
Track	1.90 m (6.23 ft.)	1.90 m (6.23 ft.)	4.80 m (15.75 ft.)	—
Empty Weight	1,182 kg (2,606 lb.)	1,185 kg (2,612 lb.)	2,700 kg (5,952 lb.)	—
Loaded Weight	1,412 kg (3,113 lb.)	1,375 kg (3,031 lb.)	3,150 kg (6,945 lb.)	—
Maximum Speed	166 kmh (103.1 mph)	180 kmh (111.8 mph)	150 kmh (93.2 mph)	—
Climb, 1000m	5.8 min	4 min	9 min	—
Climb, 4000m	48 min	—	—	—
Armament	2 fixed machine guns, 4 small bombs	2 fixed machine guns, 4 small bombs	2 flexible machine guns, 2 flexible 20mm Becker cannon	1 fixed and 1 flexible machine gun, 1 flexible 20mm Becker cannon

J-Type Order and Production Summary

AEG

- About 135 AEG J.I aircraft were built; the rest were AEG J.II, making the J.II the most numerous J-type.
- 508 AEG J-types were built through October 1918, and 609 were built through December 1918.

Albatros

- 125 Albatros J.I were ordered and delivered.
- 150 Albatros J.II were ordered and most, perhaps all, were delivered.

Junkers

- All Junkers J.I serial numbers are documented, but only 227 aircraft were delivered of the 283 ordered.
- 100 Junkers J.I aircraft were ordered from Linke-Hofmann as the Junkers J.I(Li), but none were delivered. The order may have been cancelled.

Notes

- All J-type aircraft were produced by the company that designed them. In the case of the Junkers J.I, this became the Junkers-Fokker Company.
- After a gap in serial numbers, the next serial numbers actually used always started with the number J.X00; for example, J.400/17 and J.100/18.
- It is possible that prototype J-types had serials in the ranges listed as not used in 1917.
- The AEG J.III prototype was either J.375/18 or J.575/18.
- The AEG J.II series J.375–574/18 is conjectural. It is included because of the 609 AEG J-types built by the end of 1918; otherwise there are not enough serial numbers for these aircraft. Also, AEG J.II J.417/18 was used as a civil aircraft in 1919, proving that some aircraft in this serial number range were built.

J-Type Production Orders					
Serial Numbers	**Qty**	**Not Used**	**Type and Order Date**	**Lowest Known Serial**	**Highest Known Serial**
1917 Serials					
J.1–99/17		99	Serial numbers not used	—	—
J.100–149/17	50		Junkers J.I ordered 2-19-17	J.100/17	J.149/17
J.150–159/17	10		AEG J.I ordered March 1917	J.150/17	J.156/17
J.160–199/17		40	Serial numbers not used	—	—
J.200–249/17	50		AEG J.I ordered May 1917	J.203/17	J.246/17
J.250–399/17		150	Serial numbers not used	—	—
J.400–424/17	25		Albatros J.I	J.400/17	J.421/17
J.425–427/17	3		Junkers J.I ordered 11-3-16	J.425/17	J.427/17
J.428–527/17	100		AEG J.I & J.II ordered July 1917	J.466/17	J.527/17
J.528–699/17		172	Serial numbers not used	—	—
J.700–774/17	75		Albatros J.I	J.701/17	J.758/17
J.775–799/17		25	Serial numbers not used	—	—
J.800–899/17	100		Junkers J.I ordered 11-6-17	J.800/17	J.899/17
1917 Subtotal	413	486			
1918 Serials					
J.1–99/18		99	Serial numbers not used	—	—
J.100–124/18	25		Albatros J.I	J.101/18	—
J.125–174/18	50		Albatros J.II ordered Feb. 1918	J.126/18	J169/18
J.175–374/18	200		AEG J.II ordered March 1918	J.191/18	J.374/18
J.375–574/18	200		AEG J.II (see notes below)	J.375/18	J.439/18
J.575/18	1		AEG J.III Prototype?	—	—
J.576–615/18	40		Junkers J.I ordered 6-20-18	J.576/18	J.615/18
J.616–715/18	100		Albatros J.II	J.616/18	J.714/18
J.716–725/18	10		Junkers J.I ordered 7-12-18	J.716/18	J.725/18
J.726–805/18	80		Junkers J.I	J.726/18	J.805/18
J.806–825/18	20		Ju. J.I, cancelled in favor of 20 Ju. CL.I	J.806/18	J.825/18
1918 Subtotal	705	99			

The best reconstruction of J-type production orders possible to make at this time is given above. Many thanks to Reinhard Zankl whose research enabled him to create this table and who generously provided it for this book.

When reviewing this table, it is important to know that the German authorities deliberately left gaps in the serial numbers to reduce the intelligence information the Allies could derive by statistically analyzing the serial numbers of aircraft brought down in Allied lines. Factory markings analysis, as it is now known, can provide information on the production quantities of weapons and components; leaving gaps in the serial numbers reduces the accuracy of factory markings analysis.

AEG J-Types

Above: The AEG J.I J.153/17 in factory sprayed camouflage finish at Adlershof for type testing clearly shows its derivation from the earlier AEG C.IV. Changes to the C.IV to create the J.I included replacing the 160 hp Mercedes D.III with the more powerful 200 hp Benz Bz.IV, addition of armor plate around the cockpit and engine, elimination of the fixed, forward-firing gun for the pilot, and addition of ailerons to the lower wings for improved maneuverability.

AEG J.I

The successful AEG C.IV, with its steel-tube airframe, was well-suited as a starting point from which to develop an armored J-type. To develop the new type quickly, flat armored sheets, three on each side and three on the bottom, were bolted directly onto the metal airframe of the C.IV. The armor weighed a total of 400 kg. Integrating the armor plates as structural elements would have saved some airframe weight but cost a lot of time to develop and manufacture, and the AEG J.I was needed at the front as soon as practical. To improve maneuverability with the greater weight, ailerons were added to the lower wings. Finally, to haul the extra weight into the air, a 200 hp Benz Bz.IV engine replaced the 160 hp Mercedes D.III engine used in the C.IV. At first the wing area was unchanged from the C.IV, and the additional weight raised the stall speed and lengthened take-off runs.

Assembly of the AEG J.I prototype began in January 1917 and was completed in April. The first five production J.I aircraft were delivered in May. On 30 May 1917, AEG test pilot Theodor Schauenburg was killed when he attempted to loop a J.I and the wings came off. What possessed him to perform this terminally-foolish maneuver is unknown because aerobatics understandably were not a requirement for J-types.

The J.I airframe was identical to the AEG C.IV that had already passed its type-test, so *Idflieg* delayed the static-load test (on J.150/17) until August 1917 to expedite deliveries to the front. This was long after the J.I had entered combat, which was unusual for German aircraft. Starting in September 1917 the first running changes were introduced into J.I production to decrease vulnerability to ground fire. These changes were installation of double lift-bracing wires and double control cables. The AEG J.Ia engineering prototype was used to evaluate additional changes that were introduced as production progressed (see table on page 11).

Because the AEG J-types were both relatively easy to built as well as resistant to ground fire, the AEG J.I and J.II became the most numerous J-types at the front. Like other J-types, the AEGs were assigned in ones or twos to divisional *Flieger Abteilungen* to fly contact patrol sorties, or '*Iffeln*.' In the field, the AEG J.I was complimented for its high speed, estimated to be 150–200 km/h in level flight and as high as 250 km/h in a shallow dive. Unfortunately, this was optimistic; actual top speed in level flight

Right: Front view of AEG J.I J.153/17 showing the factory sprayed camouflage and the lead for the wireless antenna hanging under the fuselage. Ailerons on all four wings provide added maneuverability to the heavy aircraft, but no horn balances are fitted to this early-production J.I.

Below: Early production AEG J.I J.151/17 at *FA(A)* 250.

Below: AEG J.I J.220/17 *Margo* of *Flieger Abteilung (A)* 208 wears the unit's reverse swastika marking.

Above: AEG J.I of *Schlasta* 26b wears that unit's red flames marking on its nose.

Above: AEG J.I J.236/17 wears sprayed camouflage with late German insignia and has two downward-firing guns.

Above: *Flieger Abteilung (A)* 208 lineup at Voyenne in the Spring of 1918 shows a variety of two-seaters with that unit's reverse swastika marking on their fuselages, including an AEG J.I at center.

Below: A closeup of the above photo of *Flieger Abteilung (A)* 208 shows the AEG J.I at right, with a captured Breguet 14 at left. Was the AEG J.I at right *Margo* as shown in the photo on the previous page?

was a mere 140 km/h. Due to its greater wing area and lighter weight, the Albatros J.I had a shorter take-off run, enabling it to operate from muddy airfields. In contrast, the heavy AEG J.I required a longer take-off run from drier surfaces. Though reasonably fast for an armored two-seater, it was not a fast airplane. Furthermore, the AEG J.I was cumbersome and difficult to land because the short fuselage and heavy nose made it easy to flip over on landing. On the other hand, crews felt more secure in the AEG J.I than the Albatros J.I because, unlike the Albatros, the engine was protected by armor. The armor's effectiveness was demonstrated by British firing tests on the armor plate of a captured J.I; results are given in the adjacent table.

Installation of a 20mm Becker cannon in an AEG J.I fitted with an enlarged gun ring was undertaken in January 1918, followed by firing trials in February–March. The firing trials demonstrated the cannon was difficult to operate in the air as mounted, making this type of installation unsatisfactory for two-seaters.

AEG J-Type Modifications

AEG introduced running changes into production of their J-types; the key changes are summarized here.

Changes from AEG C.IV to Create AEG J.I

- Change engine from 160 hp Mercedes D.III to 200 hp Benz Bz.IV
- Ailerons added to lower wings for better maneuverability
- Armor added to protect engine and crew
- Pilot's fixed gun deleted

Running Changes Introduced During AEG J.I Production

- Double lift-bracing wires and double control cables installed to increase resistance to ground fire (tested on AEG J.Ia)
- Two fixed, downward-firing guns added for more offensive capability (also used in AEG J.II)
- Lengthened fuselage for improved stability, especially during landing
- Enlarged wing area for shorter takeoff run by increasing wing chord (tested on AEG J.Ia)
- Improved aileron control linkage for better maneuverability (tested on AEG J.Ia)

Changes Introduced to Create AEG J.II

- Front spar composed of two adjacent tubes to reduce susceptibility to ground fire
- Horn-balanced ailerons (upper wings) for improved controllability (various configurations used)

Changes Introduced During AEG J.II Production

- Wing spars, fuselage frames, and aileron control horns strengthened
- Horn balances added to elevators and rudder for improved maneuverability
- 20 J.II aircraft built with 20mm Becker in floor, Sept. 1918
- All J.II aircraft had two downward-firing guns except those with 20mm Becker in floor

Changes Introduced to Create AEG J.III

- 260 hp Mercedes D.IVa replaced 200 hp Benz Bz.IV for better performance
- Redesigned, lightened armor plate for better performance

Results of British Testing on AEG J.I Armor Plate Effectiveness

Ammunition	Angle to Normal, °	Safe Range, Yards	Unsafe Range, Yards
German A.P.	0	—	600
	15	500	400
	30	400	300
Mark VII.P Armor Piercing (angle to normal approximate)	0	700	600
	15	400	300
	30	300	200
German Spitze	0	150	100
	15	100	50
	30	50	50
Mark VII	0	50	—
	15	50	—
	30	50	—

British authorities performed an engineering analysis of an AEG J.I brought down on 16 May 1918. During the course of the analysis, the ability of various types of ammunition to penetrate the 5mm thick armor at various ranges and angles was tested, resulting in the data tabulated above. The armor was judged to be too light to protect against British armor piercing ammunition below 500 feet altitude.

Above: Typical early production AEG J.I with wireless antenna lead visible below the observer's cockpit.

Former observer Hanns-Gerd Rabe flew the AEG J.I in combat. "In order to make close observation possible and to take part in the land battles, armored airplanes were built with the engine, fuel tank, and pilot well-protected by armor plate. The observer was less protected because he had to move about freely to make his observations. Equally unprotected were the radiator above the engine and the control cables. At the end of July 1918 my unit, *Flieger-Abteilung (A)* 253, received an armored airplane of the AEG J.I type. We regarded it with suspicion for it was too heavy and seemed to be clumsy as the 200 hp engine was too weak. The pilots agonizingly circled the airfield with it, but it barely went over 500 meters. Risky maneuvers in aerial combat were out of the question... In our opinion it was a very primitive bird!" This was the opinion of many combat aircrewmen, who clearly did not agree with Idflieg's theories on required J-type performance.

Continuing Rabe's comments, "The take-off was laborious; the crate lifted off very sluggishly and hung in the air... I heard the machine gun bullets hitting the armor plating and also saw holes being torn in the fabric... but I wasn't really concerned... When I had exposed all my photographic plates, I threw my hand grenades from the lowest altitude into the English posts and dropped gas bombs before the entrances of the dug-outs, and then fired my machine guns. I could work both machine guns in the floor with a foot pedal and I also fired the rotating machine gun at the machine gun emplacements in the trenches that were mainly firing at me from the rear... Suddenly, hot water hit me in the face. The vulnerable radiator had been hit. We could not last for more than a few minutes before the engine over-heated and froze up. Peter Johannes (the pilot) turned the AEG in a flat curve... to take us back to our own lines. But in the banked position, hits on the control cables to the rudder and ailerons tore them apart and caused the machine to dive towards a thick mass of barbed wire... Of course, the 'Emil' (nickname given to German two-seater pilots; the observer was 'Franz') succeeded in once again laboriously bringing the AEG back on an even keel, but the landing gear got caught in the barbed wire and the machine rolled over, leaving the landing gear pointed skyward..."

According to a French intelligence bulletin dated November–December 1918, the fuselage of an AEG J.I 468/17 (*lang*) was found near Strasbourg. The J.I was prone to landing accidents (flipping over) because of its short fuselage and heavy nose armor, and lengthening the fuselage for better pitch stability and control was an appropriate modification. The number of AEG J.I aircraft manufactured with longer fuselages is not known.

Above: AEG J.I J.203/17 had the doubled lift wires introduced into production in September 1917 to improve survivability. Identification streamers, probably in red but black was also used, attached to both rear inner interplane struts, are hard to see because the wind is whipping them.

AEG J.I J.203/17

AEG J.I of *Schlasta* 26b

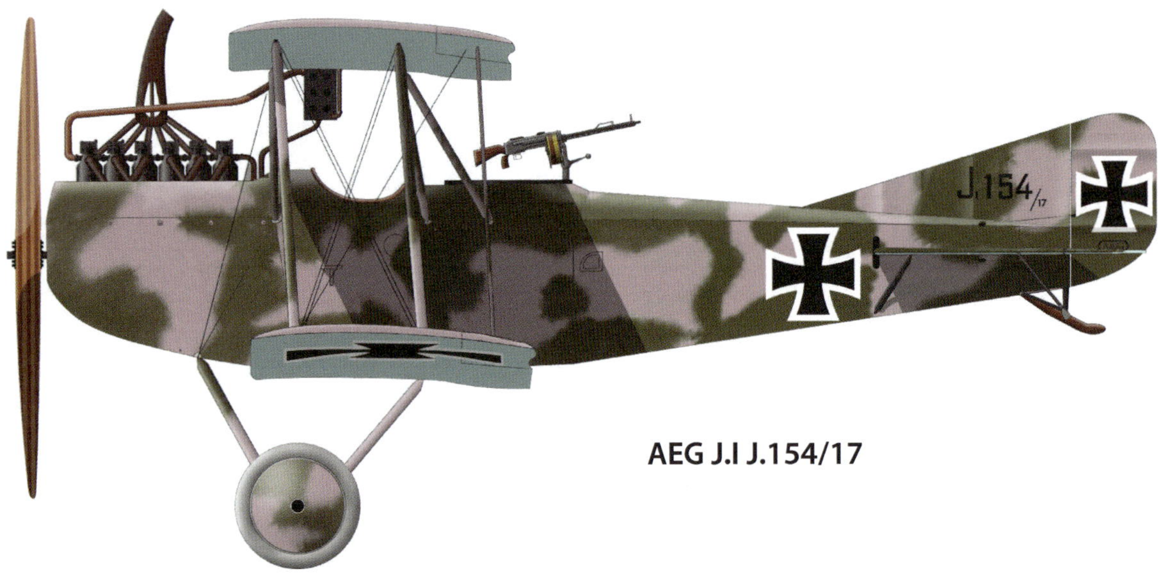

AEG J.I J.154/17

AEG J.I J.220/17 *Margo* of *Flieger-Abteilung (A)* 208

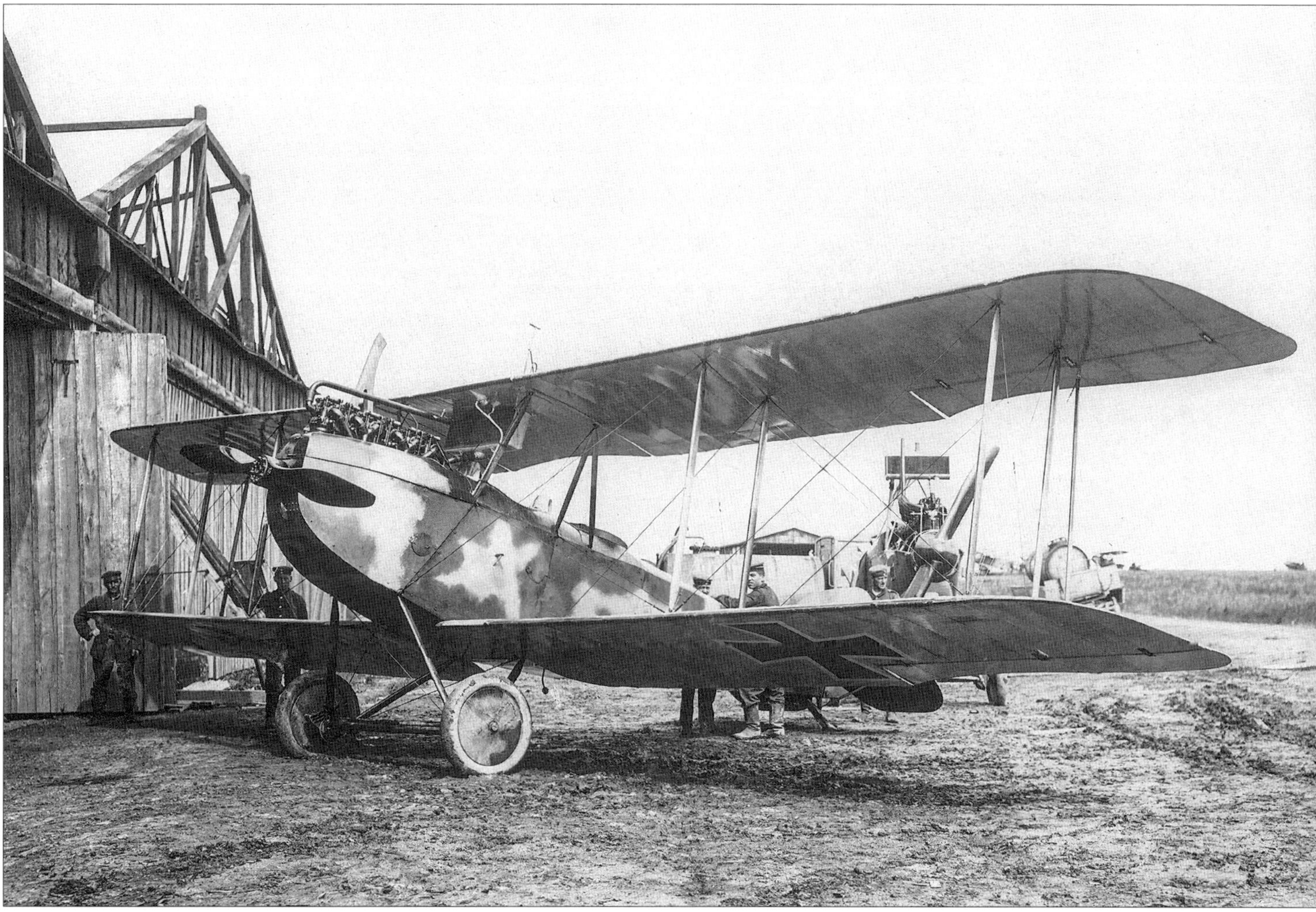

Above & Below: Two views of early-production AEG J.I J.154/17 wearing its sprayed camouflage pattern.

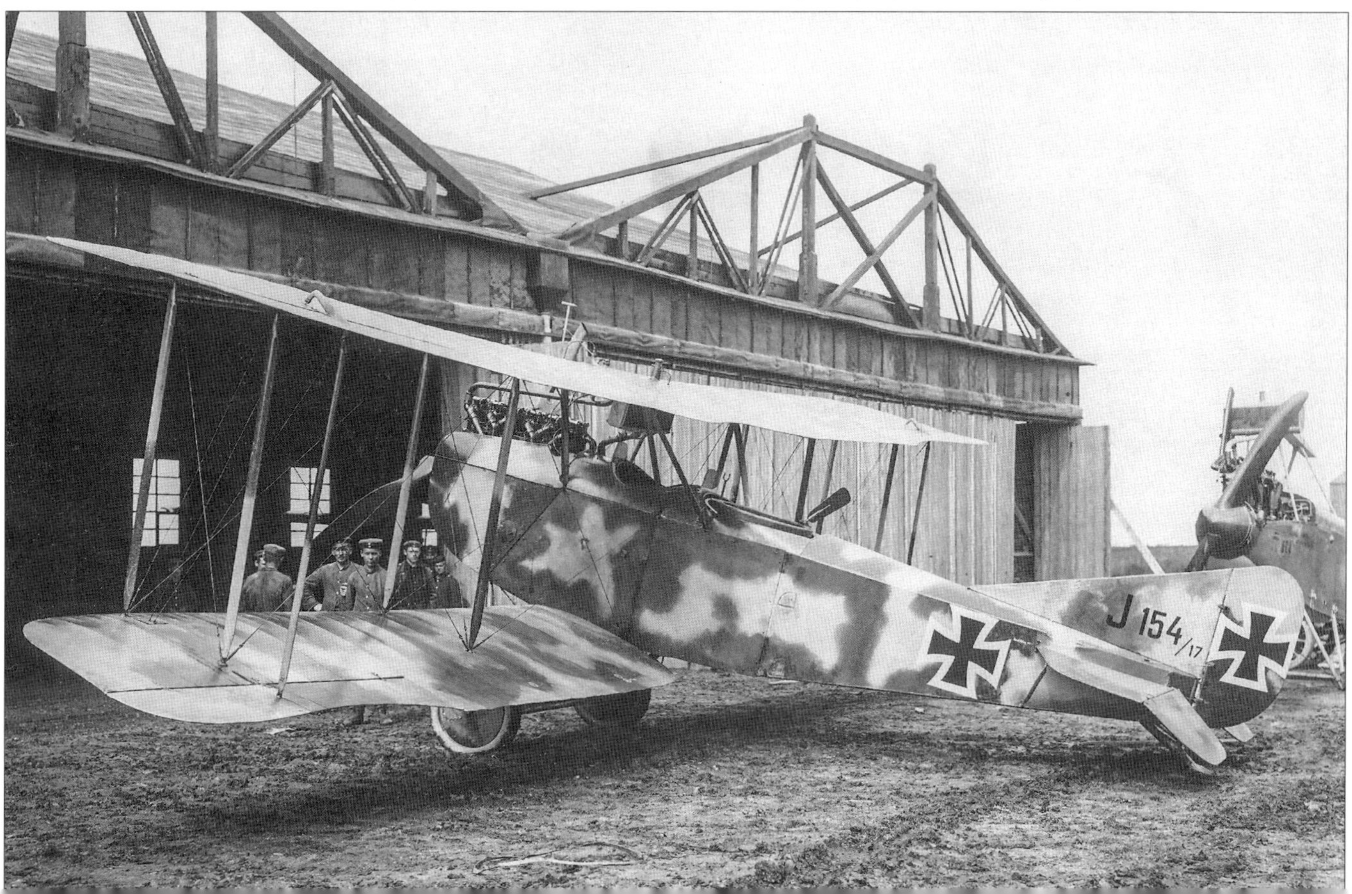

AEG J.I J.216/17 of *Flieger-Abteilung (A)* 221

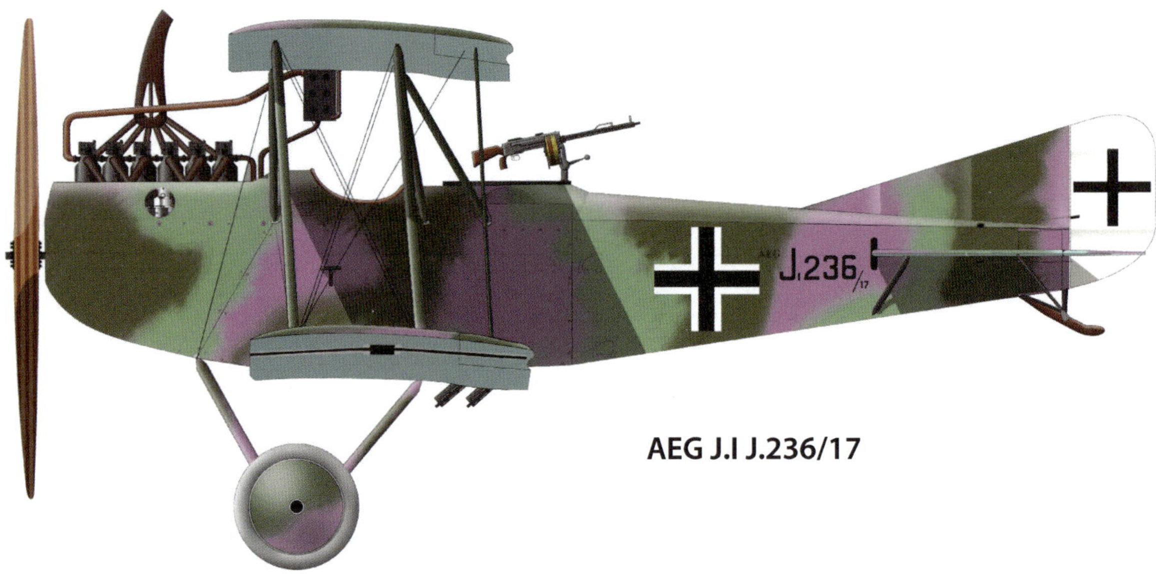

AEG J.I J.236/17

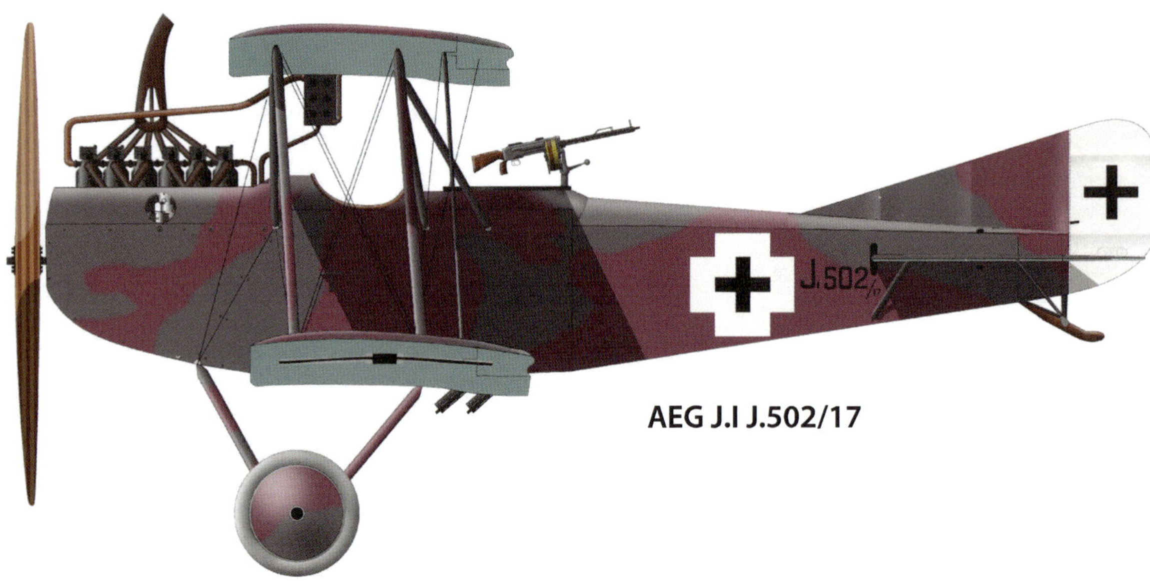

AEG J.I J.502/17

Above: Early-production AEG J.I J.158/17 in sprayed camouflage finish photographed at the AEG company airfield at Nieder-Neuendorf. The attachment to the gun ring is to counterbalance the drag of the machine gun, reducing the force the gunner needed to apply to turn the gun against the slipstream.

Above: An unidentifed AEG J.I.

Above & Below: Two views of early-production AEG J.I J.159/17 at *Flieger Abteilung (A)* 255 wearing its factory-applied sprayed camouflage pattern but no unit or individual markings.

Above: Later-production AEG J.I J.502/17 with downward-firing guns, straight-sided insignia, and later camouflage. This aircraft has the relocated aileron connecting struts as tested on the AEG J.Ia.

Above: Early-production AEG J.I with early sprayed camouflage pattern at *Flieger Abteilung (A)* 250. Red or black identification streamers are folded on the lower wing.

Above: The AEG J.Ia was a developmental aircraft to test engineering changes for the AEG J.I. Among the changes was enlarged wing area achieved by use of a greater wing chord, modification of the aileron control linkage for better maneuverability, and double lift wires and control cables to make the J.I less susceptible to fatal damage from ground fire. All these changes were introduced to the AEG J.I during its production run and were continued in the AEG J.II.

AEG engineers began work in August 1917 to improve the J.I design. The wing loading was reduced by enlarging the wing area by 12 percent to reduce the stall speed and shorten the take-off run. The wing span remained constant; the wing area was enlarged by increasing the wing chord from 1.65 meters to 1.85 meters. To reduce the pilot's control forces, which improved maneuverability, the strut connecting the upper and lower ailerons was moved and attached to the aileron linkage. Installing double lift wires and control cables made the aircraft less vulnerable to ground fire. Some of these modifications, which were tested in August–October 1917 on the AEG J.Ia, were gradually introduced on the J.I production line, and later on all of the J.II machines. AEG J.Ia was not a military designation and appears to have been an internal AEG designation for a single development aircraft.

AEG J.II

As flight testing of the AEG J.Ia continued, the design of the AEG J.II, which incorporated many J.Ia features, was nearing completion. To further reduce vulnerability to ground fire the J.II's front spar was composed of two parallel steel tubes. By October 1917 the J.II prototype was undergoing flight evaluation. In January 1918 the flight-test results were still unsatisfactory, especially the poor rate of climb. A new control wheel for the pilot gave control forces that were much too heavy, hindering maneuverability.

The evolution of J.II control surfaces during its production run is evidence that flying qualities needed improvement, but production could not be interrupted because aircraft were needed at the front. The first J.II production machines were delivered with unbalanced control surfaces. Flight testing of aerodynamically-balanced controls was especially time-consuming because it was totally empirical. Furthermore, flight testing was delayed by winter weather that limited test-flight opportunities. A few early production aircraft featured balanced ailerons within the upper wing planform, but the standard

Above: AEG J.II J.186/18 is an early production J.II without balanced control surfaces, making it look like a late-production J.I. The aircraft did have the larger wing chord and pair of fixed, downard-firing guns standard in the J.II.

Above: AEG J.II J.232/18 is another early production J.II with unbalanced control surfaces, making it look like a late-production J.I. Like the forward fuselage, the underwing radiator was armored.

AEG J.II J.186/18

AEG J.II J.20x/18

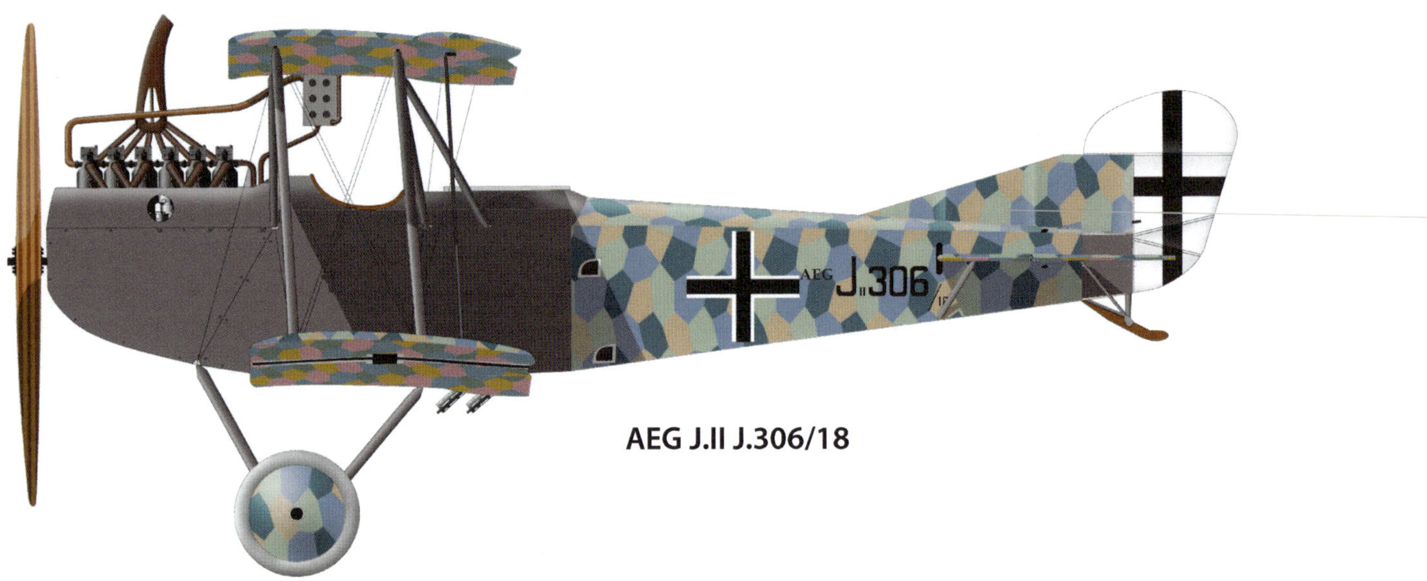

AEG J.II J.306/18

Above: Factory photo of AEG J.II J.20x/18 with ailerons fitted with inset aerodynamic balances on upper and lower wings. Balanced tail surfaces are also fitted. The pair of fixed, downard-firing guns are clearly visible.

Above & Below: An AEG J.II, possibly J.294/18, at Adlershof in August 1918. This represents the final production configuration with all control surfaces fitted with horn aerodynamic balances except the ailerons on the lower wings. The pair of fixed, downard-firing guns are visible, and all fabric surfaces are covered with five-color printed camouflage fabric.

Above: AEG J.II J.306/18 is a in final production configuration and covered with four-color hexagonal camouflage fabric overall except for the fuselage armor, which is painted a solid dark color.

configuration had projecting horn-balances for all control surfaces except the lower wing ailerons.

The AEG J.II was in quantity production by March 1918, and by April it was in combat at the front. Because of its similarity to the J.I, the static load test was apparently waived to expedite deliveries. According to combat reports, the AEG J-types performed very well during the Allied summer offensives. However, the long take-off run was still criticized. In fact, *Idflieg* considered installing the 260 hp Mercedes D.IVa engine to improve take-off performance, something that should have been done in the beginning. However, no changes were introduced, probably due to limited supplies of this popular engine. In fact, failure to employ the powerful 260 hp Mercedes D.IVa is inexplicable except as a consequence of inadequate supplies.

Several J.II crashes compelled *Idflieg* to perform a full static load test, which was done using J.II 294/18 during 8–20 August 1918. Testing revealed that in a shallow dive (Case C) the wing spar joints failed at 64% of the required load. Some fuselage frames and the aileron controls also needed strengthening. The problems were traced to a combination of poor materials, an increasing problem throughout the aviation industry at this stage of the war, and deficient manufacturing processes.

Apparently all J.II aircraft were delivered with the two-gun battery mounted in the observer's cockpit except for those with a 20mm Becker cannon. In June 1918, a single machine-gun was experimentally fitted in the nose so the pilot flying at 100 meters could fire at ground targets about 460 meters ahead. However, this was not placed in production. Also in June 1918, a flexible 20mm Becker cannon installed in the floor of the observer's cockpit was tested. Slotted armor grating protected the gun opening and aiming window. In September 1918, AEG delivered 20 J.II aircraft armed with the 20mm Becker cannon, primarily for destroying tanks.

Because of production with no recorded break between types, the exact number of AEG J.I and J.II aircraft built is unknown, but about 135 J.I and the remainder J.II is a reasonable estimate. Up to October 1918 some 508 J-types were delivered by AEG, more than any other manufacturer, which agrees with the Inter-Allied report that a total of 609 AEG J-types were built, a total that includes the machines completed in November and December 1918 and later adapted for civil use.

AEG J.III

The AEG J.III prototype, designed in September 1918, was finally fitted with the more powerful 260 hp Mercedes D.IVa engine needed from the beginning. *Idflieg* finally accepted what the aircrews knew from the beginning, that more power was needed not only to reduce take-off distance but also to increase speed and improve climb rate and agility. Greater power would improve operational flexibility, overall handling, and stall margin (for flight safety), and the ability to defend against fighter attack. In addition to the increased power, the J.III had redesigned armor plate that saved 100 kg of weight. As far as is known, the J.III did not reach series production before the Armistice and no photographs have been found.

Facing Page: Armor plate has been removed from AEG J.II J.241/18 to show its pair of fixed Spandau machine guns mounted in the rear cockpit. The guns were offset to the left to leave space for the gunner's seat on the right.

Above: The running engine of this late-production AEG J.II does not seem to disturb the crewmen in the photo.

Above: Late-production AEG J.II J.270/18 has come to grief, apparently the result of a bad landing at its aerodrome. The pair of downward-firing machine guns is visible protruding through the armor plate.

AEG J.II J.232/18

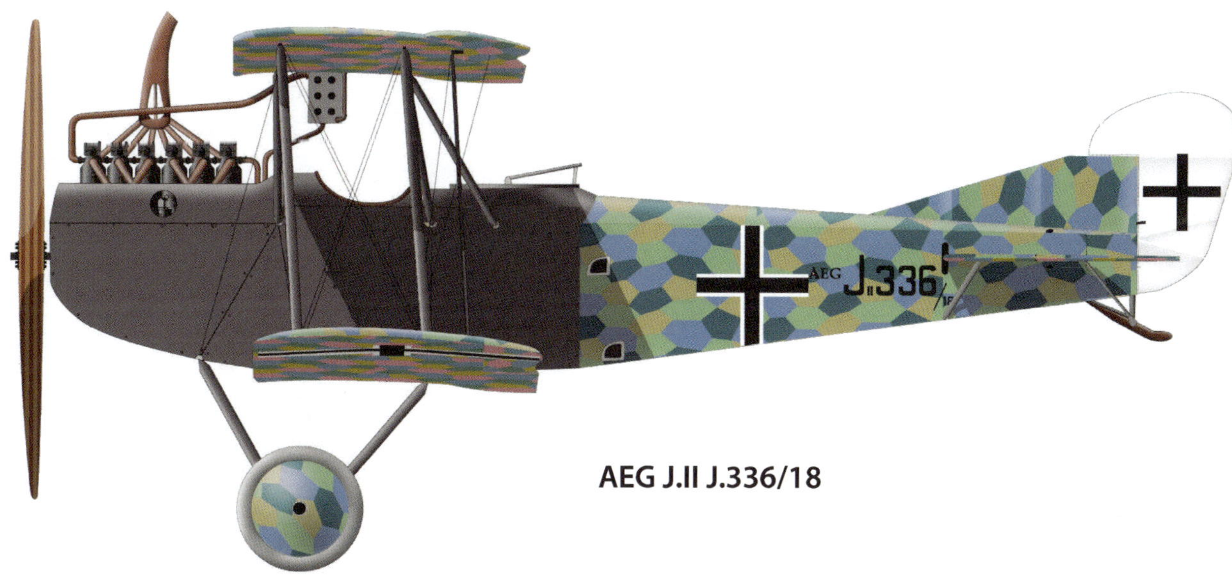

AEG J.II J.336/18

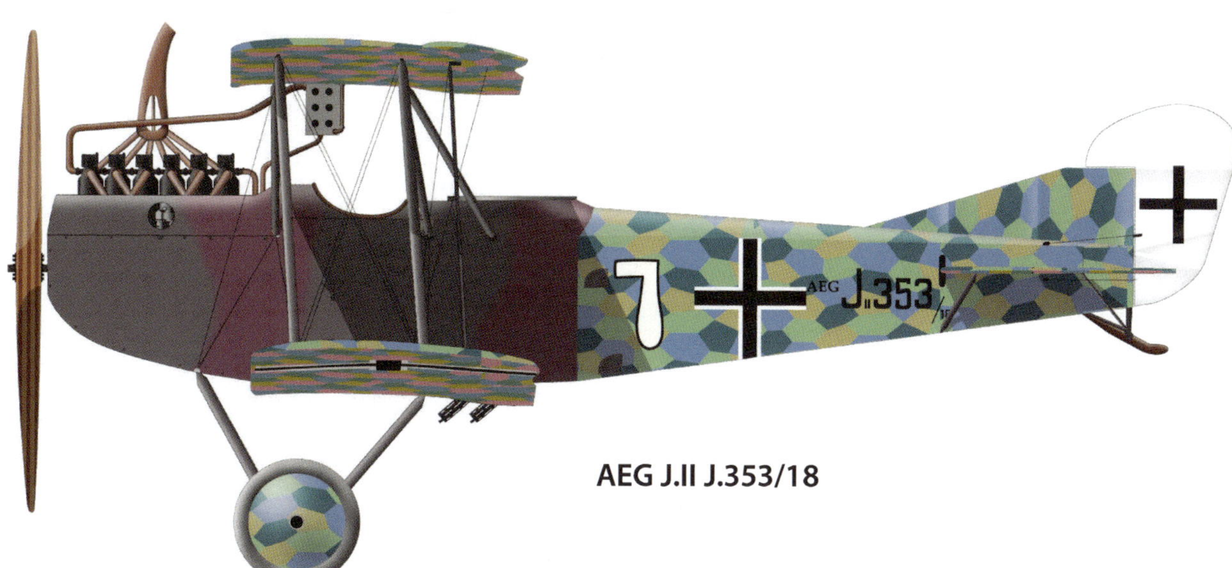

AEG J.II J.353/18

Above & Right: Late-production AEG J.II J.336/18 is shown in standard factory camouflage and markings.

Below: Mounting for twin downward-firing guns in AEG J.I and J.II.

Below Right: Observer's triggers to fire the two downward-firing guns in the AEG J.I and J.II. Both drawings from *The Engineer* of Sept. 6, 1918.

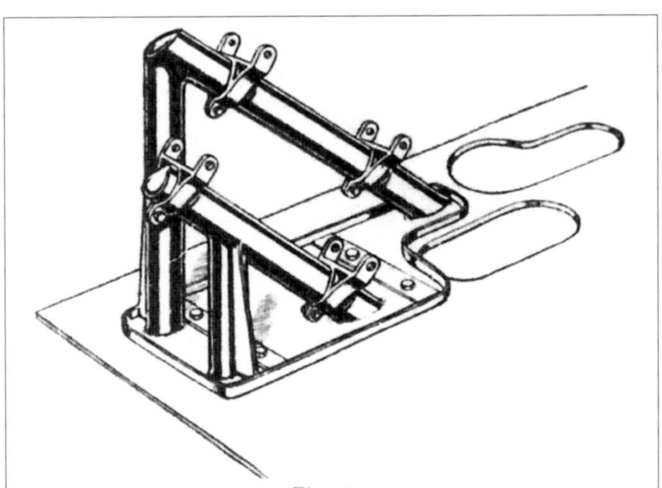

Fig. 8.

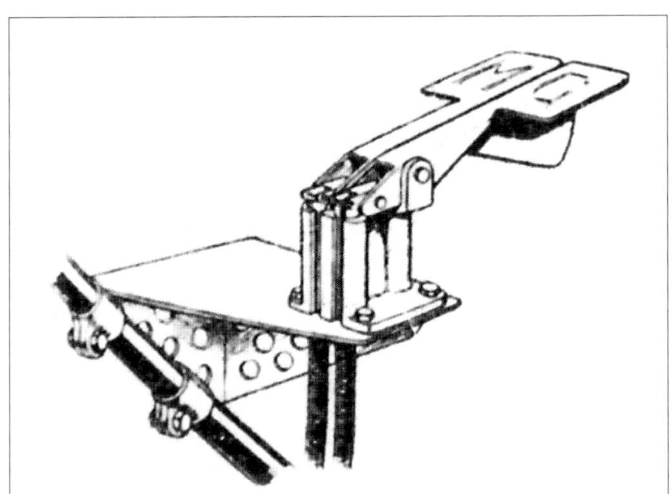

Above: Late-production AEG J.II J.336/18 is shown in standard factory camouflage and markings. Unusually, the downward-firing guns are not present.

Right: The wreck of AEG J.II J.249/18 brought down in US lines on 27 October 1918. The location was said to be ½ mile southeast of Cunel. Details of the armor are visible, including where it was bolted to the airframe. There were three plates on each side and another three under the fuselage. This photo shows the downward-firing machine guns clearly. The hole in the armor just aft of the starboard undercarriage strut enabled the gunner to view the ground beneath the aircraft and fire when the guns were pointing at a target. Arrangements for the downward-firing guns in the AEG and Albatros J-types were similar.

Facing Page, Top: The fuselage of AEG J.II J.353/18 is shown in the huge storage facility in Romorantin, France, where many German aircraft were stored post-war. It was tactical number '7' in its unit. AEG J.II 353/18 with the white "7" on the fuselage is also visible in photos of the Zeppelin hangar at Trier (Treves in English and French) too, indicating that it was turned over to the Allies at Trier, Germany, then transported to Romorantin, France.

This Page: This is a series of photos of the wreck of AEG J.II J.249/18 brought down in US lines on 27 October 1918. The location was said to be ½ mile southeast of Cunel; the captions on the photos indicate it was downed by a fighter. Details of the armor are visible and the bottom photo shows the downward-firing machine guns.

Below: Postwar photo of German warplanes gathered outside a Zeppelin hangar at Trier, Germany. Many German aircraft were collected there for distribution to the Allies under terms of the armistice.

Right: AEG J.II in a photo of German aircraft outside the Zeppelin hangar at Trier postwar; it is a detail of the photo below.

Above: An AEG J.II is at right in this postwar lineup, with two LVG C.VI aircraft at left.

Above & Left: A AEG J.II '13' serving with the postwar D.L.R. airline before and after a landing accident. The J.II still wears its wartime camouflage and markings with D.L.R insignia and markings superimposed. Armor was removed from civil AEG J.II aircraft.

Above, Left, & Below: Three different AEG J.II aircraft serving with the postwar D.L.R. airline. *Deutsche Luft-Reederei*, founded in 1917, was the first German airline to use heavier than air aircraft. D.L.R.'s first air services began on 5 February 1919. The company used a stylised crane, designed by Professor Otto Firle, as its logo.

Above: A civil AEG J.II with passenger cabin postwar; unfortunately, the men are not known.

Below: A civil AEG J.II with passenger cabin serving with D.L.R. postwar. In 1919, D.L.R. was one of the founding members of International Air Traffic Association, the predecessor to today's IATA.

In 1923 D.L.R. merged to became part of Aero Lloyd AG, who again merged to became part of Deutsche Luft Hansa in 1926, which also adopted the crane logo.

The Albatros J-Types

Above: Albatros J.I J.401/17, the second production aircraft, is shown at Adlershof undergoing evaluation. The nose has a spinner and streamlined shape because there was no armor around the engine. The Albatros J.I was the only J-type without engine armor and was therefore the aircraft most vulnerable to ground-fire of all the J-types.

Albatros J.I

The initial Albatros armored design, the Albatros J.I (company designation L40), used the same basic wood wings and tail surfaces as the preceding Albatros C.XII reconnaissance plane. These readily-available components were coupled with a new engine and forward fuselage featuring 5 mm steel sheet armor around the crew cockpits. To maintain center of gravity with the lighter engine, the wings were given 2° of sweepback. In contrast to the airfoil radiator of the Albatros C.XII, the J.I had a box radiator installed in front of the leading edge of the upper wing. To limit weight, the J.I engine cowling and radiator were left un-armored; this led to frequent engine failures due to damage from ground fire.

Like all production J-types, the Albatros J.I used the Benz Bz.IV engine. This 200 hp engine was substantially less powerful than the 260 hp Mercedes D.IVa used in the C.XII from which the J.I was derived. The reduced power, coupled with the increased weight of the armor and increased drag, significantly reduced J.I performance and agility compared to its reconnaissance aircraft predecessors. Initially the Albatros J.I had a very streamlined nose complete with propeller spinner, but this was soon found to be unnecessary for such a slow aircraft and later production aircraft eliminated the spinner, which actually reduced drag.

Initially, the armament consisted of one flexible Parabellum MG14/17 for the observer. Later, some aircraft were equipped with two LMG 08/15 machine guns fixed in the fuselage floor, which could fire diagonally forward and down at a 45° angle. Furthermore, the crews carried light bombs and grenades, which were thrown by hand to attack troops and other ground targets.

The Albatros J.I was the first German aircraft that deployed a cannon in unrestricted operational service, as distinct from operational trials. In November 1917 the first tests began with the 20 mm Becker cannon, which was installed on the left side of the observer's cockpit of Albatros J.I J.710/17.

After firing 600 rounds in ground tests with the

Above: Albatros J.I J.401/17 at Adlershof. The J.I used the wings and tail of the Albatros C.XII fitted to its armored fuselage. The fuselage behind the armor was wood like the Albatros C-types. The wireless antenna lead is hanging below the observer's cockpit.

Above: An early production Albatros J.I, perhaps J.401/17, showing the propeller spinner used on the early machines.

machine's rear fuselage elevated, in-flight firing tests began from the Albatros on 12 December 1917. It was shown that the recoil of the cannon proved substantially less than during the ground tests. Moreover, installation of the cannon hardly affected the in-flight handling characteristics. Target acquisition was deemed good, but the ammunition clip proved difficult to change in the slipstream. By the end of January 1918 testing was successfully completed with the exception of the clip-loading problem. In February this appeared to be solved by provision of a special handgrip on the ammunition clip. Three cannon-armed J.I aircraft were then sent to the front, primarily intended for tank busting, followed by five more in April. Each of these machines had a captured Lewis machine gun installed in the observer's cockpit for defense against fighters; the Lewis was used to save weight.

Above & Below: Albatros J.I J.415/17 with factory finish and *Flieger-Abteilung (A)* 238 unit markings. The early J.I aircraft used the wing and tailplane camouflage of the late Albatros C-types like the C.XII together with varnished wood rear fuselage and light-gray armor. Over this J.415/17 has black and white elevator stripes, white stripes over the upper wing, and a white field on the upper fuselage behind the gunner with a Roman numeral painted on it, in this case 'VI', which were the *Flieger-Abteilung (A)* 238 unit markings. The closeup view shows the armored door to the pilot's cockpit open.

Above: Albatros J.I J.421/17 serving with *FA(A)* 293; the older style iron cross has been removed and the new style straight cross painted forward. The spinner and painted upper wing camouflage are first production batch features.

Above: Albatros J.I J.400/17 was sold to Austria-Hungary and became 09.01. Here it is ready for transport and may be on its way to Austria because the spinner has been removed and 09.01 had no spinner.

These eight cannon aircraft underwent combat evaluation with *Schlachtstaffeln* 10, 17, and 28b during May–June 1918, producing mixed results. *Lt.* Umlauff, the *Idflieg* cannon expert who visited various units to demonstrate the new weapon, flew Albatros J.I 734/17 on six missions over the front, but was severely wounded by an enemy fighter while attached to *Schlachtstaffel* 17. The "expedient" fuselage mounting was criticized because the side location restricted the field of fire, which depended too much on aircraft position and attitude in relation to the target. In addition, it was impossible to repair stoppages aloft. *Schlachtstaffel* 28b, experiencing problems with stoppages and handling the 15-round magazine, was convinced that a valuable weapon would result from adoption of belt-fed ammunition.

Albatros J.1 Production Batches

Albatros J.I production orders totaled 125 aircraft in three batches. Review of available photographs appears to show these characteristics:

First Batch of 25, serials J.400–424/17:
Spinners, painted camouflage on top of wings and tail surfaces (like the Albatros C.XII).

Second Batch of 75, serials J.700–774/17:
No spinners, painted camouflage on top of wings and tail surfaces (like the Albatros C.XII).

Third Batch of 25, serials J.100–124/18:
No spinners, printed camouflage fabric on all flying surfaces. Some had patterned camouflage on fuselage.

Note: All Albatros J.II aircraft were apparently finished like late-production J.I aircraft.

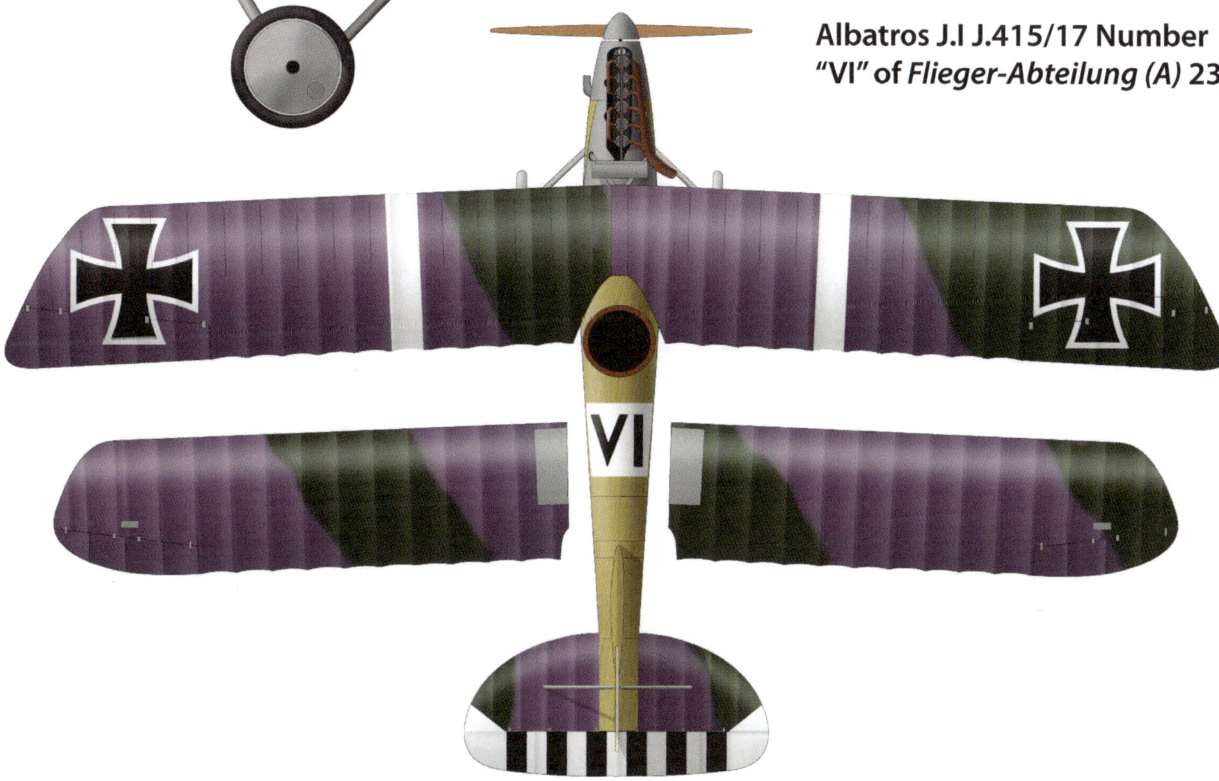

Albatros J.I J.415/17 Number "VI" of *Flieger-Abteilung (A)* 238

Albatros J.I J.421/17 of *Flieger-Abteilung (A)* 293

Above: This Albatros J.I may be from the second production batch; it has no spinner but does not appear to have printed camouflage fabric. It has been painted overall dark camouflage colors, probably green and lilac like other Albatros two-seaters, with white nose, nose stripe, and letter 'L', probably at *Flieger-Abteilung (A)* 233 where it was serving when photographed.

Below: Closeup of an Albatros J.I radiator reveals that it was not armored. Together with its totally exposed location, that made it especially vulnerable to ground-fire. Engine failure could be expected two–three minutes after radiator puncture. Use of a box radiator instead of an airfoil radiator like that of the C.XII did nothing to improve performance of the J.I.

Above: Early production Albatros J.I with spinner, perhaps at Adlershof. Wind tunnel investigations eventually revealed that eliminating the spinner and rounding the nose actually reduced drag, and subsequent production J.I aircraft had no spinner. The same change was made to more well-known aircraft like the Rumpler C.IV and late-production Albatros D.III fighters built in Austria-Hungary.

Above: Closeup of an Albatros J.I showing details of the armor, the wood rear fuselage, and the gunner with his flexible machine gun. The J.I had little of the grace of the elegant Albatros C.XII from which it was derived.

Above & Left: Some Albatros J.I aircraft mounted a 20mm Becker cannon on the left side of the aircraft for ground attack, and especially destroying tanks, as shown here.

Schlachtstaffeln 10 reported no gun stoppages, but noted the ammunition clip did not function smoothly and the side mounting restricted the field of fire. Moreover, accurate aiming was difficult because the tracer and explosive ammunition exhibited different ballistic characteristics. The *Schlachtstaffeln* expressed the unanimous opinion that the effectiveness of cannon-armed ground-attack aircraft depended wholly on keeping enemy fighters at bay.

Kofl 6 reported that in 12 missions the cannon had proved extremely effective, especially in dawn or dusk attacks on rear-area transportation, and more cannon-armed aircraft were urgently requested. For destroying tanks it was suggested that armor-piercing ammunition be supplied. The recommendation to install the cannon on a pivot mount in the cockpit floor was accepted and tested in at least one Albatros J.I. On 11 June 1918, the first Albatros J.I with a Becker cannon, J.768/17, was forced down in enemy lines by ground fire.

In December 1917, three Albatros J.I without engines were purchased by the *k.u.k. Luftfahrtruppe*,

Above: Albatros J.I J.738/17 of the second production batch has two-color camouflage over all upper surfaces and a white 'V' identification marking.

Above: This view of an early-production Albatros J.I shows the upper surface camouflage pattern.

as Series 09, and received the serial numbers 09.01–09.03. Originally, Albatros J.I 09.01 carried the German *Militär-Nummer* J.400/17; 90.02 was formerly J.726/17, and 90.03 was formerly J.730/17. Marta-built 250 hp Benz engines were installed. On 28 February 1918, aircraft 09.02 crashed in flames during evaluation tests at Aspern. On 5 May 1918, aircraft 09.01 was damaged at the front and returned to Aspern for repairs. The only J.I used operationally by the *Luftfahrtruppe* was 09.03, which served with *Flik* 69/S between July and September 1918.

Below: Albatros J.I after a bad landing.

Above: This Albatros J.I serving with *FA* 33 is likely from the second production batch.

Below: The fuselage of Albatros J.I J.707/17 is moved the old-fashioned army way, soldier power.

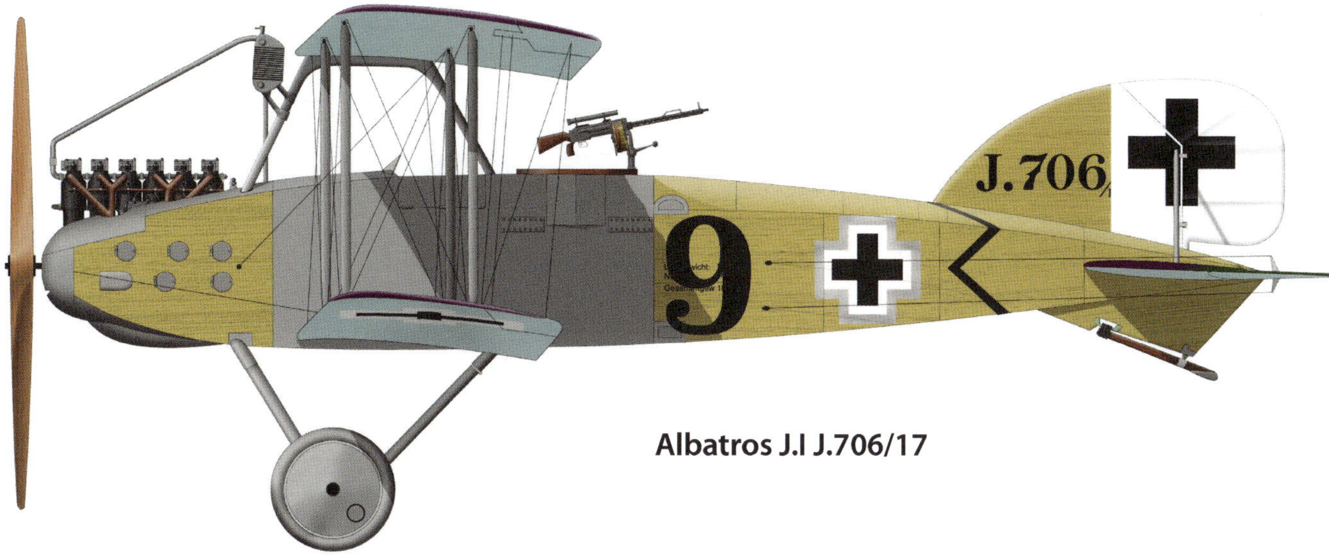

Albatros J.I J.706/17

Albatros J.I J.738/17

Albatros J.I "L" of *Flieger-Abteilung (A)* 233

Above: Albatros J.I J.706/17 is from the second production batch. In addition to the factory finish it carries a dark, probably black, zig-zag marking on the rear fuselage and tactical number '9'.

Above: This Albatros J.I is likely from the second production batch.

Below: This Albatros J.I is likely from the second production batch and may be the same aircraft in the photo above.

Above: This Albatros J.I, tactical letter 'H', has experienced a bad landing. Downward-firing guns were installed.

Right: The cockpits of an Albatros J.I; the pilot's control wheel and side doors of armor are visible along with some details of the observer's cockpit.

Above: Albatros J.I J.727/17 of the second production batch wears only factory finish and markings.

Below: A derelict Albatros J.I appears to be the victim of shelling or a bombing raid.

Above: Albatros J.I J.758/17 is shown with crew in this excellent portrait. Did 'Teddy' (rear inboard strut) fly any missions?

Above: Albatros J.I J.756/17 poses with ground crew. Two downward-firing guns are visible.

Left: Albatros J.I poses with its flight crew.

Above: Colorized photo of Albatros J.I J.758/17 with crew in this excellent portrait courtesy of Jim Miller.

Albatros J.I J.758/17

Above: As war reparations several German aircraft, including at least one Albatros J.I, were sent to Japan, where a contemporary artist painted details of their camouflage, which are reproduced above for the J.I.

Above, Above Left, & Lef: These three photos of the demise of Albatros J.I J.769/17 were found in an album from Spa 150, suggesting a Spad from that unit downed the J.I and its unfortunate crew.

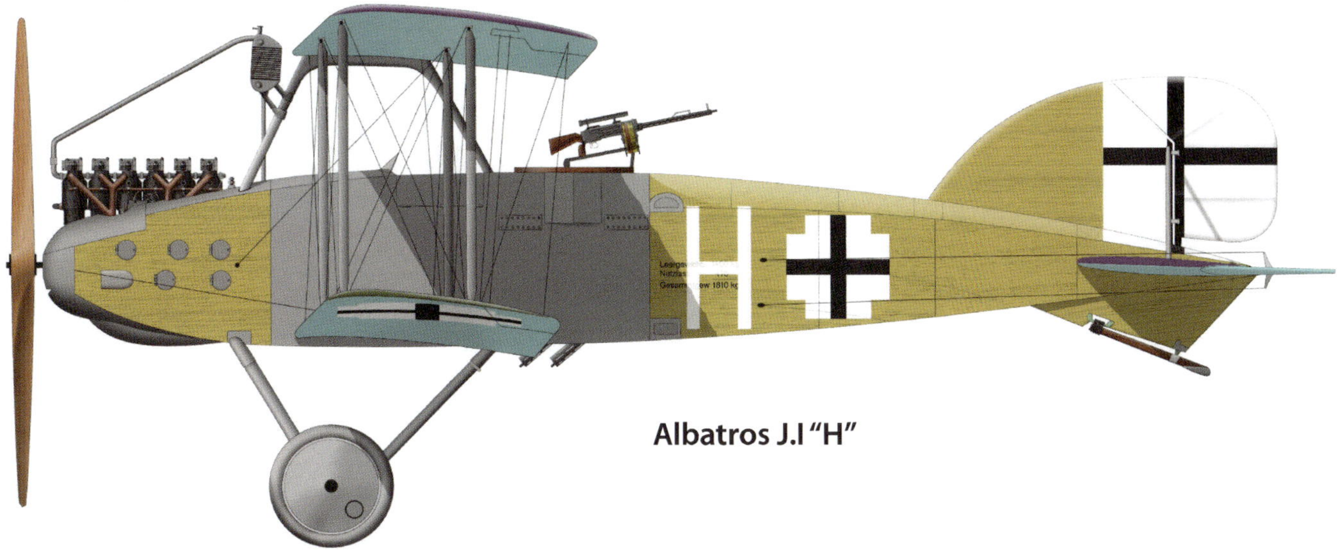

Albatros J.I "H"

Albatros J.I J.769/17

Albatros J.I 09.01 in Austro-Hungarian service; formerly J.400/17

Above & Below: Albatros J.I 09.03 was the last of three purchased by Austria-Hungary. Formerly J.730/17, 09.03 was the only J.I used operationally by the *Luftfahrtruppe;* it served with *Flik* 69/S between July and September 1918. The two-color camouflage painted on the upper surface of the wings and tailplane is evident below.

Above: Albatros J.I 09.01 was formerly German *Militär-Nummer* J.400/17. Marta-built 250 hp Benz engines were installed. On 5 May 1918 Albatros J.I 09.01 was damaged at the front before seeing operations and returned to Aspern for repairs. It no longer has the propeller spinner it had when it was built.

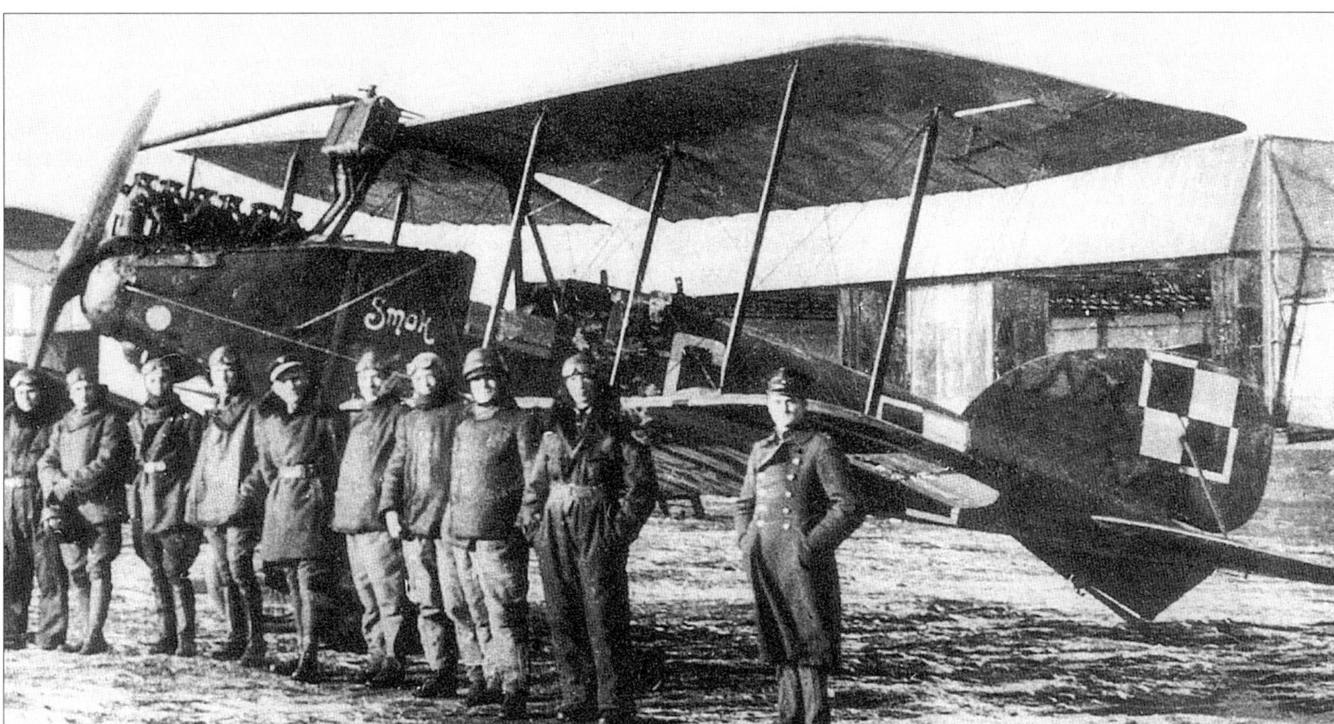

Above: Post-war the new Polish state used a number of German warplanes, including the Albatros J.I. As can be seen on its fin, this J.I, named *Smok* and with tactical number '6', wears the irregular camouflage pattern illustrated by the Japanese artist in color on the top of page 50.

Above & Left: Polish Albatros J.I tactical number '6', apparently before it was named. The photo above shows the irregular camouflage pattern illustrated by the Japanese artist in color on the top of page 50 on its fuselage. The photo at left is too dark to show the pattern clearly but does show the printed camouflage fabric on the wings and horizontal tail surfaces.

Albatros J.I "6" *Smok*

Above: The Albatros J.II was redesigned to improve survivability in the hostile environment in which it flew its missions. Most importantly, the entire engine was now armored, solving the worst problem with the J.I. In addition, unlike the J.I, the radiator was now armored, further improving survivability. Finally, the upper ailerons were given aerodynamic horn balances to reduce control pressures for better maneuverability. The same basic Benz engine was used, but power was boosted 10% to 220 hp to compensate for the greater weight of the engine armor, which accounted for the boxy cowling. The result was a more robust, survivable aircraft with essentially the same performance despite its greater weight.

Albatros J.II

Although the Albatros J.I met the letter of the J-type requirement for armor protection of the crew, it was the only production J-type that did not have an armored engine. Unsurprisingly, lack of engine armor was the flyers' main complaint about the Albatros J.I, leading directly to the redesigned J.II. Like the competing Junkers and AEG designs, the Albatros J.II featured armor around the engine, giving the J.II a boxy nose instead of the streamlined nose of the Albatros J.I. In addition, the radiator was likewise armored to make it less vulnerable.

To compensate for the additional weight of the engine armor, the J.II had the 220 hp Benz Bz.IVa, a slightly more powerful version of the 200 hp Benz Bz.IV engine installed in the J.I. Wing sweep was reduced slightly to maintain proper center of gravity with the additional nose armor. Despite the additional power, maximum speed of the J.II remained the same as the J.I due to the increase in weight and drag – a slow 140 km/h. Like the preceding Albatros J.I, climb and ceiling of the J.II remained modest. Nevertheless, since most missions were flown at 500 meters or less, climb rate and ceiling were not viewed as critical for J-type aircraft, which relied on their armor for protection instead of speed and altitude performance.

Like the preceding C.XII and J.I, the J.II had ailerons on all four wings. However, to reduce control forces and improve responsiveness, the upper wings of the J.II were given horn-balanced ailerons. Tail surfaces remained the same as the J.I. Initially, armament was also the same; a flexible machine gun for the observer and two fixed machine guns firing downward at a 45° angle. However, the Albatros J.II was designed to carry a flexible 20mm Becker cannon mounted in the floor of the observer's cockpit. But *Idflieg* reported that, by the end of September 1918, no J.II aircraft carrying this armament had been delivered. Forward-firing armament had not been installed on J-type aircraft because it was not considered useful for air-to-air combat given the heavy J-type's lack of agility. Likewise, diving at ground targets for strafing was at first considered too risky for these heavy, armored aircraft. However, due to difficulty aiming the downward-firing guns, as J.II production progressed the two fixed guns reportedly were moved to the conventional position in front of the pilot and firing directly ahead. If so, it was the only J-type to use this armament configuration. Unfortunately, photographs to confirm this installation have not surfaced.

While the Albatros J.II had better armor protection than the J.I, its greater weight with essentially the same wing further increased its stall speed and aggravated the take-off distance, agility, and flight safety challenges suffered by all J-types. Albatros J.II production orders totaled 150 aircraft.

Above Right: The 20mm Becker cannons fitted to the Albatros J.Is proved to be effective anti-tank weapons, and 20 Albatros J.II aircraft were fitted with these weapons in an improved mount in the floor of the gunner's cockpit shown here; the front of the aircraft is at the top of the photo. Extra magazines for the Becker are stacked on the sides of the armored cockpit and extra drums of machine gun ammunition are stored at the rear. Twenty AEG J.II aircraft were delivered with a similar Becker installation.

Below: Albatros J.II 127/18 is shown here is its basic factory finish and markings. Flying surfaces are covered with printed camouflage fabric, the armor is painted light gray, and the wood rear fuselage and fin are clear-varnished.

Above: Albatros J.II 140/18 serves in basic factory finish and markings; no unit or personal markings are visible.

Below: The Albatros J.II, with its armored engine and radiator, was much more likely to survive being hit by rifle-caliber ground fire than its J.I predecessor.

Above: Some Albatros J.II aircraft had the fuselage camouflage seen on some J.I aircraft and illustrated in color by the Japanese artist on page 50. No unit or personal markings are visible; the tires are missing due to the rubber shortage.

Below: This postwar view of German aircraft sent to Italy as reparations shows an Albatros J.II in the middle with the same pattern of fuselage camouflage seen above, another indication it was applied at the factory.

Above: Albatros J.II 126/18 serves in basic factory finish and markings; no unit or personal markings are visible.

Below: This rear view of the Albatros J.II displays its standard factory finish and markings. Printed camouflage fabric covers all flying surfaces, the armor is painted light gray, and the wood rear fuselage and fin are varnished.

Above: Albatros J.II ready for a mission clearly showing the downward-firing guns and observer's gun. The wind-driven dynamo for the wireless is on the front undercarriage strut, and the leads for the antenna are below the fuselage.

Below: Albatros J.II after a landing accident; the downward-firing guns are barely visible under the fuselage.

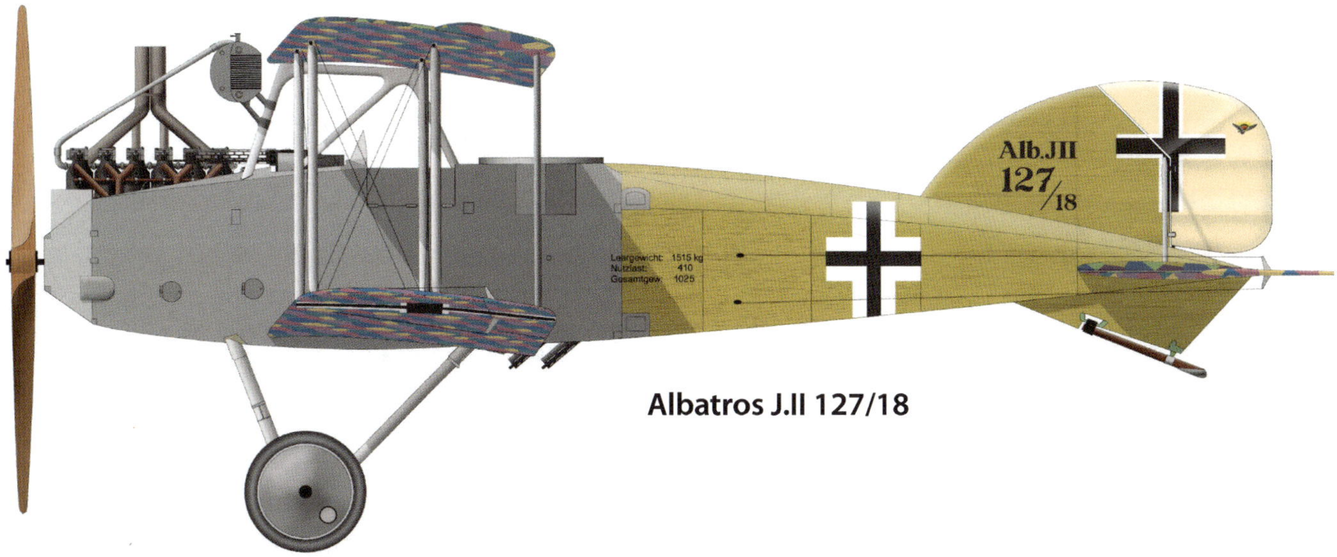

Albatros J.II 127/18

Albatros J.II, serial unknown

Above: Albatros J.II ready for a mission.

The Junkers J.I

Above: Junkers J.I J.101/17, the second production aircraft, was sent to *Flieger Abteilung (A)* 250 for front-line evaluation. It is seen here in basic factory finish and markings; no unit or personal markings are visible. The cantilever all-metal wings required no bracing wires, a drag-reducing feature that contributed to the Junkers J.I being the fastest J-type.

Professor Hugo Junkers met with *Hauptmann* Muhlig-Hofmann of *Idflieg* on 26 October 1916 to discuss the forthcoming armored aircraft specification. Although Junkers proposed a parasol monoplane, *Idflieg* insisted on a biplane, likely due to distrust of monoplanes' structural strength. On 3 November 1916, *Idflieg* ordered three Junkers all-metal, armored biplane prototypes (J-type) made of Duralumin(1). Albatros and AEG received similar contracts for J-type aircraft simultaneously.

The Junkers J.I was designed by *Doctor-Ingenieur* Otto Mader, *Diplom-Ingenieur* Franz Brandenburg, and *Ober-Ingenieur* Steudel. Shortly before roll out, the Junkers J.I prototype (designated J 4 within the company) was inspected by *General* Ernst von Hoeppner, commander of the *Luftstreitkräfte*, and *Major* Wagenführ of *Idflieg*. To show Hoeppner the superiority of metal aircraft, several conventional wood-and-fabric aircraft were set on fire by incendiary bullets.

Before armor was installed, company pilot Arwed von Schmidt made a few brief flights in the Junkers J.I prototype at Adlershof on 27 January 1917. The next day the machine was moved to the proving ground at Doberitz where Schmidt performed its first real flight. Taking off in snow 20cm deep, Schmidt reported a take-off run of 200 meters. During this flight an altitude of 250 meters and a top speed of 145 km/h were reached. The engine ran rough and Schmidt landed after a mere 3½ minutes. Schmidt found the prototype to be steady and stable, and, lacking its armor plate in the forward fuselage, it was tail heavy as expected. Schmidt reported the elevator control was insufficient but rudder response was excellent and the unbalanced ailerons operated

(1) Duralumin (also called duraluminum, duraluminium or dural) is the trade name of one of the earliest types of age-hardenable aluminium alloys. Duralumin was developed by the German metallurgist Alfred Wilm at Dürener Metallwerke Aktien Gesellschaft starting in 1903 and leading to the introduction of duralumin in 1909. It was used for rigid airships and, starting in 1916, for Junkers aircraft.

Above: The prototype Junkers J.I photographed January 28, 1917 at Döberitz airfield. The military number of this aircraft is unknown. The prototype had no horn balances on rudder or ailerons. Armor was not installed, so it was tail heavy.

Below: *Diplom-Ingenieur* Franz Brandenburg poses at Döberitz airfield on May 3, 1917 with the Junkers J.I that he helped design. The rudder and fin have been enlarged from the prototype above and the rudder now has a horn balance to reduce the pilot's control forces, but the ailerons do not. The exhaust manifold differs from that of production aircraft.

Above: The same Junkers J.I that was photographed on May 3, 1917, was photographed again on May 23, 1917. Now the elevators are dural; earlier they were covered with fabric. The fin is now braced by two struts, and a pitot tube is on the upper wing for flight testing. The aileron horn balances used on production aircraft are not fitted to this aircraft.

Below: Junkers J.I J.102/17 has come to grief in a landing accident on September 10, 1917, a relatively common experience for all aircraft at the time. More significantly, despite the great weight of the aircraft, it did not collapse on the aircrew. The third production aircraft, it has horn balances on the ailerons. It was repaired and sent to the front the week of January 28, 1918.

Above: Junkers J.I J.118/17, an early production aircraft, serving with of *Flieger Abteilung (A)* 208. Despite its lack of colorful markings, the Junkers J.I was held in great regard. "On 28 March 1918, I piloted your machine J.128 on an infantry flight at 80 meters. We received over 100 hits, my crash helmet was pierced, and the observer received a grazing wound. I am convinced that your machine save my life. With no other machine it is possible today to fly over the enemy trenches at such low altitude." So wrote a grateful *Lt.* Wagner of *Flieger Abteilung (A)* 268 to Hugo Junkers.

Below: Junkers J.I J.138/17 is seen here with the armored engine access door open while undergoing maintenance. The first 40 production Junkers J.I aircraft had fabric-covered rudders, but beginning with aircraft J.140/17 an all-metal rudder was standardized. J.I J.140/17 was shipped to an *Armee Flugpark* during the week of March 11, 1918 for acceptance testing.

Above: Junkers J.I J.109/17 serves as a backdrop for a unit portrait. Interestingly, this early production aircraft has survived long enough at the front in its hazardous duty to have the new national insignia of March 1918 applied.

easily. He regarded the landing characteristics as acceptable.

Because flight tests were encouraging, *Idflieg* placed an order for 100 Junkers J.I aircraft on February 19, 1917. The first production batch was for 50 J.Is, numbered J.100– J.149/17 and powered, like competing J-types, by the 200 hp Benz Bz.IV. *Idflieg* was concerned about the lack of mass-production experience at Junkers & Co. (Ico), and subsequent experience was to show the concern was well placed. Obtaining the special armor plate was another challenge because production was delayed. On 24 May 1917 the armor plate producer, Hüttenwerk Dillingen, finally reported the armor plate met requirements and production deliveries could start.

The Junkers cantilever wings did not have a conventional spar. Instead, the wings were built up from a number of dural tubes spaced around the aerofoil section.[2] Because stresses gradually decreased from the wing roots to the tips, the tubes were deliberately reduced in size toward the tips by inserting smaller sections like a telescope; this reduced weight. These span-wise tubes were interconnected with riveted bracing made from short dural strips or tubing to stiffen the wing structure. The corrugated wing skin, made of 0.015 inch thick dural attached by rivets, created a strong, stiff wing without using conventional ribs.

The fuselage was also built of dural tubing. The pre-assembled armor tub, which weighed 470 kg and protected the engine, fuel tank, and crew, was part of the forward fuselage structure to minimize weight. The rear fuselage was covered with fabric and the horizontal stabilizers and elevators were dural. Up to aircraft J.139/17 the rudder and the fin were fabric-covered dural. Beginning with aircraft 850/17, the entire empennage, including fin and rudder, were all-metal. The wing and undercarriage struts were made of steel tubing bolted to steel lugs riveted to the armor plate. The struts were covered with dural fairings for streamlining. The engine bearers were made of wood instead of metal to minimize passing engine vibrations to the airframe.

Idflieg began the required *Typenprüfung* (type test) at Adlershof in early May 1917 with a flight evaluation and assessment of the airframe, engine, and instrument and equipment installations. The type-test aircraft was a prototype that did not meet official delivery regulations due to the Junkers company's inexperience in building military aircraft. Many minor details and accessories were missing, and more important, the observer's machine-gun was reported as being useless because the upper wing's overhang prevented the observer from standing upright, making it impossible for him to properly aim his gun. The design also made it difficult for a

Above: Junkers J.I J.594/18 and J.596/18 on the Junkers airfield at Dessau; they were sent to the front in September 1918. The nearest aircraft is still waiting for its rear fuselage to be covered with printed, camouflage fabric.

tall pilot to enter the cockpit. Flight characteristics, especially maneuverability in roll that was restricted by the long wing-span, were unsatisfactory. Due to the type's potential, the type-test committee decided to release the prototype for combat evaluation and wait until a production aircraft could be examined to make a final determination on the J.I's suitability for production.

Static load-tests were performed on 19 and 21 May and 4 and 5 June for Case A (leveling out from a dive) and Case B (gliding flight). Testing for dive (Case C) and inverted flight (Case D) were not required for J-types. Although the wing easily exceeded its load requirements, the rear fuselage failed a load test on 14 July and, despite being reinforced, failed again on 24 July 1917. Likewise, the elevator had to be reinforced. Meanwhile another prototype was undergoing flight evaluation and modification at Doberitz, while production aircraft were being assembled at Dessau.

The observer was armed with the standard observer's armament for German two-seaters, a Parabellum LMG 14 machine gun in a rotating ring turret. Three ammunition drums were handy on the right and left of the seat, and two additional drums were stored under the seat, giving a total of 1,000 rounds. Per J-type requirements, no forward-firing armament was installed.

Two machine guns, fixed to fire downward at 45° angle, were installed in the floor of the observer's cockpit of J.816/17 in January 1918. The camera had to be removed for the guns to fit. The downward-

Above: Junkers J.I J.104/17 has suffered a landing accident, a common occurrence for all WWI airplanes on the rough fields of the time.

(2) For additional information about Junkers metal aircraft, see Peter M. Grosz, *Junkers D.I*, Windsock Datafile No.33 and Jan Servaites, "The Junkers Monoplanes and Metal Shaping Techniques," *Over the Front* Vol.20 No.1, Spring 2005.

firing guns added unwelcome additional weight to an already heavy, under-powered aircraft, and the installation was not adopted for production. A flexible 20mm Becker cannon was also installed in a Junkers J.I in January 1918 and flight firing trials were flown in early February. The Becker cannon was difficult to handle in flight with the mounting it had, making it unacceptable for combat. Therefore the Becker cannon installation was not placed in production either, and the only armament carried by production Junkers J.Is was the observer's flexible machine gun.

Combat Experience

The first production machine, J.I 100/17, was shipped to Flanders (4th Army) to join *Flieger-Abteilung* 19. That unit introduced the J.I to combat on 1 August 1917. Later that day Junkers pilot Arwed von Schmidt, who had chaperoned the J.I to the front, sent a telegram to Junkers stating,

Above: Junkers J.I J.802/17 being held in place before take-off.

Right: Junkers J.I J.819/17 wears identification streamers attached to its lower wingtips and has no spinner.

Junkers J.I J.101/17 of *Flieger-Abteilung (A)* 250

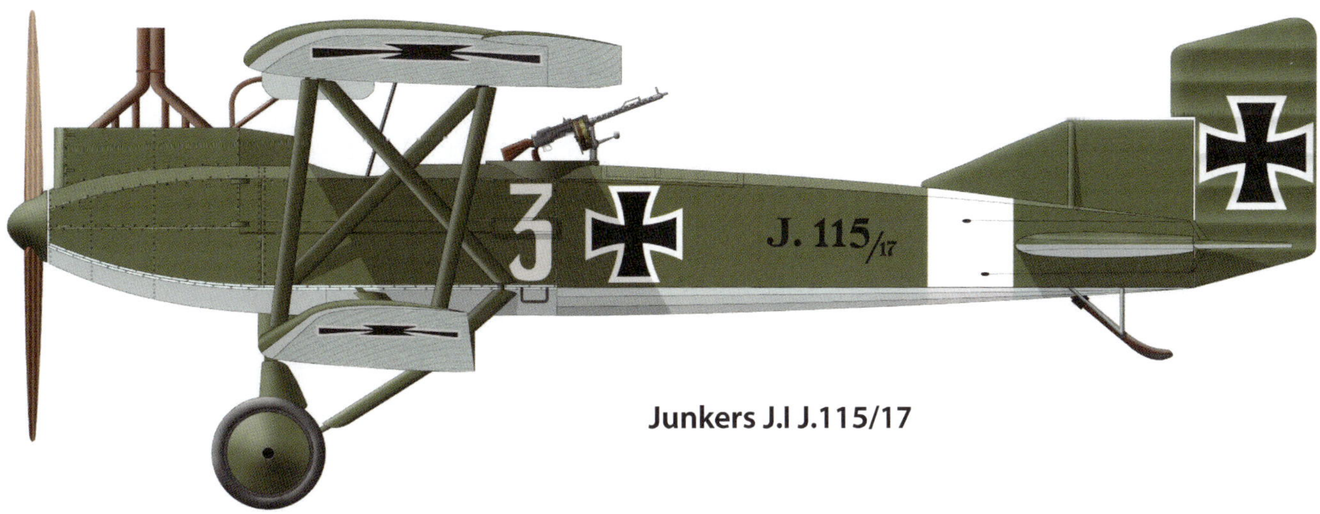

Junkers J.I J.115/17

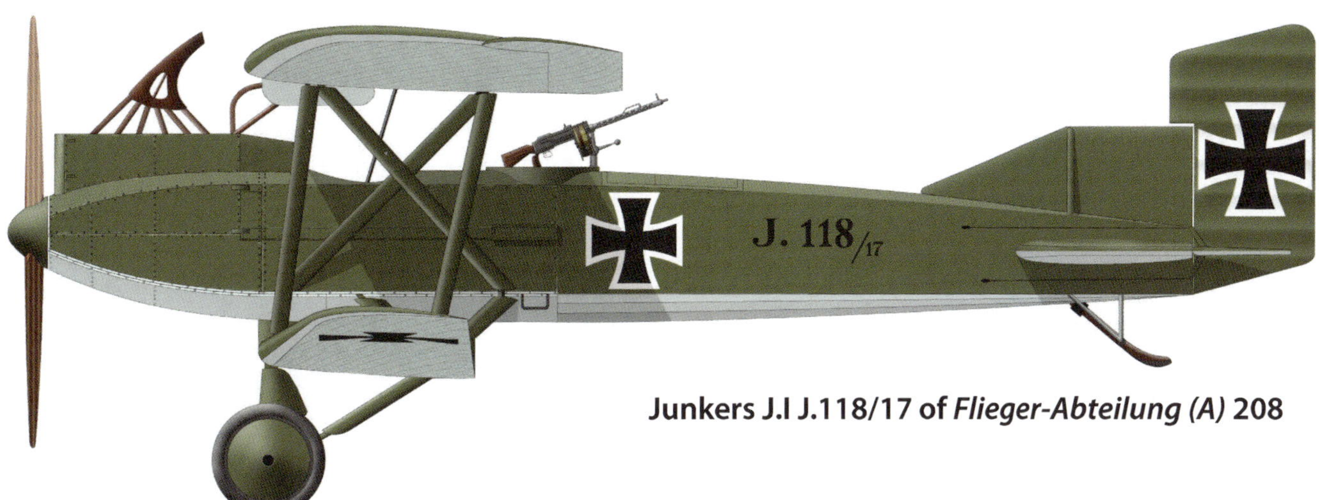

Junkers J.I J.118/17 of *Flieger-Abteilung (A)* 208

Above: Junkers J.I J.750/18 is readied for its next mission, and this time it appears it will have the luxury of a Halberstadt CL.IV escort to help defend it against enemy fighters. The photo is from the August Quoos album and may be postwar.

"the *Abteilung* commander was very appreciative of the J.100, having flown at 50 meters over Ypres and received 11 hits in the wings." *Abteilung* 19 appreciated the Junker's relative invulnerability and speed that were superior to the AEG and Albatros J.I types. Schmidt flew the machine in weather conditions that kept most aircraft on the ground, which impressed the combat aircrews. On 23 September 1917, a pilot reported that the J.100, in spite of having received 85 hits, remained in perfect condition. Minor combat damage was fairly easily repaired with a special tool kit that Junkers provided with each aircraft.

By 4 October 1917 the Junkers J.I had accumulated enough combat experience that a consensus of opinion had formed. *Flieger-Abteilung* 19 and *(A)* 250 reported that the Junkers J.I was stable and remained steady in strong wind gusts, possessed a high glide ratio, and was relatively nimble for such a heavy aircraft. They especially commended the excellent glide ratio because it enabled pilots to glide the aircraft back across the lines after engine failures, which the J.I experienced several times. Maneuverability and speed were adequate for the J-type's design role of infantry cooperation.

However, the take-off run was very long because the aircraft was so heavy, limiting it to operating from larger airfields. During German retreats in 1918 J.I aircraft sometimes had to be abandoned because they could not take off from smaller airfields in poor condition. Pilots had to maintain airspeed during landing approach to avoid excessively hard landings, which meant higher touch-down speeds than for other two-seaters. Nevertheless, pilots viewed the Junkers J.I as easier to land than the AEG J.I, which could flip over on landing.

The all-metal J.I could remain exposed to the elements for weeks without apparent damage. In combat the aircrews soon became attached to the Junkers due to the great security they felt because of its effective armor and robust airframe. For additional protection against rifle-caliber fire, many observers flew into combat wearing a standard infantry helmet.

Of course, opinions about its flying qualities differed between units. *Flieger-Abteilung (A)* 250 reported that the Junkers J.I required highly experienced pilots, whereas *Flieger-Abteilung* 19 wrote that even novice pilots found the aircraft easy to fly because it gave them so little trouble. A few pilots objected to the stick control, so one J.I was fitted with a control wheel at the Junkers airfield in Wurzen and delivered to *Idflieg* for tests on 4 April 1918. Apparently there was little or no improvement because a control stick remained standard. Aircrews nicknamed the Junkers J.I *der fliegende Tank* (the flying tank) and *der fliegende Mobelwagen* (the flying moving van) because of its great sturdiness and weight.

Like other J-types, the Junkers J.I was assigned in ones or twos to frontline units under divisional or Army command. The Junkers J.I was flown during times of active combat to spot enemy troops in no-man's land and those waiting in ready trenches up to a kilometer behind the forward trenches. German troops, often cut-off from communication with rear command posts during combat, usually displayed white identification panels on the ground to define their forward position. This was particularly important in a battle of movement that characterized many of the actions late in the war. By locating friendly troops while flying through hostile airspace, aircrews provided an important service for army commanders. One *Abteilung* wrote the Junkers

Above: Junkers J.I J.817/17, an early production aircraft, also serves in basic factory finish and markings. Despite its lack of colorful markings, the Junkers J.I was held in high regard by its crews.

Below: A Junkers J.I J.138/17 of *Flieger Abteilung (A)* 287 at Monceau le Neuf airfield.

This widely-distributed propaganda photograph shows Junkers J.I J.802/17 being loaded with food to drop to ground troops. In reality, the food and ammunition were packed in strong wicker containers to protect them when dropped.

J.I was unconditionally reliable and had gained the highest confidence of the aircrews despite increased enemy machine-gun fire. Further, its upper-surface camouflage was reported as very effective, one unit stating that even their own escort aircraft had difficulty in keeping the Junkers in sight and recommended enlarging the cross on the upper wing to help the escort aircraft maintain visual contact with the Junkers.

Enemy fighter attacks were considered the greatest threat to aircraft flying infantry patrols. For one thing, the pilot of the Junkers J.I had no upward visibility to maintain visual contact with the attacking fighter, and a single observer with his flexible gun was not considered adequate to defend against a determined attack by several fighters. Accordingly, combat units recommended the Junkers J.I be escorted by fighters, and CL-types were sometimes assigned that duty. Despite this concern, there is no proof that any Junkers J.I was shot by Allied fighters. Of course, there were losses due to ground-fire, engine failure, and pilot error; for example, J.I J.134/17 was downed when its pilot was killed when hit in the head.

Above: Junkers J.I J.858/17 awaits its next mission; it wears the standard upper-surface camouflage of dark green with mauve patterns that was effective against the ground.

Production Problems

After awarding the production contract, *Idflieg* became increasingly distressed that Junkers J.I mass production was not making enough progress. In January 1917, Junkers had projected a production rate of 30–50 J.I aircraft per month starting in April, but he had badly underestimated the complications involved in transitioning to mass production of an aircraft built with such innovative technology and materials. For some reason *Idflieg* decided the problem could be solved by combining Junkers' engineering skills with Fokker's production know-how. Perhaps *Idflieg* chose Fokker because his aircraft used welded steel tube frames, but that technology was much different than Junkers' dural designs. Regardless, *Idflieg* forced the merger of the two companies by refusing to provide further orders unless Junkers (Jco) merged with Fokker. Junkers strongly resisted a joint venture and Fokker was indifferent. Nevertheless, under *Idflieg's* pressure the Junkers-Fokker Werke AG (Ifa) was founded on 20 October 1917. In response, *Idflieg* gave the new company a production contract for 100 Junkers J.I aircraft, numbered J.800/17 to 899/17, on 6 November.

In retrospect it is obvious that such a forced collaboration could not succeed, and it should have been obvious at the time, but wishful thinking prevailed. Not only were Fokker and Junkers totally incompatible personalities, the flamboyant Fokker was hardly a production expert. In fact, Fokker's most notorious failures during the war (Fokker's Dr.I and E.V wing failures) were failures of quality control during production. Fokker's genius lay in showmanship, exceptional flying ability, design innovation, and an intuitive feel for modifying aircraft for good flying qualities, not production or quality control expertise. Moreover, the technology for producing a sophisticated all-metal aircraft was foreign to Fokker, whose own designs were made of wood, fabric, and welded steel tubing, not dural.

Unsurprisingly, Fokker quickly realized the Junkers production team were inexperienced and the factory was poorly organized. Despite investing a lot of money in the joint venture and sending a few foremen and skilled works to Junkers, Fokker himself took no part in solving Junkers' production problems; he was already fully busy with his own company. While Fokker remained disengaged, the Junkers team slowly resolved their production problems. The average monthly output in 1918 was 17 J.I aircraft, and the highest monthly output was 31 aircraft in October. The Junkers workforce, numbering 367 men and 155 women in January 1918, doubled to 724 men and 346 women by November 1918.

Bringing a new structural technology into mass production was not the only hurdle. Winter floods shut down armor production in January 1918, and no J.I aircraft were delivered in March because the Junkers factory ran out of armor plate. The Junkers factory had no company airfield at its location at Dessau, which delayed aircraft assembly and acceptance. The first J.I aircraft had to be shipped to Doberitz for assembly and acceptance, impeding production. Several unsatisfactory arrangements followed, first a move of assembly and acceptance flying to the airfield at Wurzen at the end of 1917, then, in the middle of February 1918, *Idflieg* decreed the aircraft were to be shipped directly to *Armee Flugpark* 17, a depot behind the front, for

Above: Junkers J.I J.148/17 and J.800/17 are in this lineup at the recently-completed Junkers airfield at Dessau.

acceptance. However, this was too far from the factory to work very well. Finally Junkers purchased land for an airfield near the factory, and flight testing was performed there from 1 May 1918. For some reason Junkers J.I acceptance flights were shifted back to an *Armee Flugpark* in August 1918. When J.I production ceased in January 1919, 227 Junkers J.I aircraft had been built out of a total of 283 ordered from Junkers.[3]

Design Developments

Junkers J.I J.749/17 was used for the installation of a more powerful 260 hp Mercedes D.IVa engine, which should have been used in this heavy aircraft from the beginning, but apparently was not available due to demand from other aircraft production programs. The first flight of the modified J.I was on 16 July 1918. This aircraft validated the benefit of greater power with a shorter take-off run, better rate of climb, and improved maneuverability. Regardless, production J.I aircraft continued to be underpowered by the 200 hp Benz Bz.IV engine.

On 2 October 1918, Junkers met with his staff to discuss future development of Junkers armored aircraft. Junkers opened the meeting by rejecting a step-by-step improvement of the J.I in favor of a truly significant step forward, an armored, parasol monoplane that would meet his personal aerodynamic standards for an aircraft. A parasol monoplane would give an excellent field of view downward, and the observer's field of fire and pilot's visibility upwards were to be improved by using a narrow-chord wing. Junkers seriously thought about cutting off J.I production before a new design was available because of the J.I's long take-off run.

The next day Junkers was absent while the meeting continued. His staff was more realistic, and they decided to proceed with a lightened version of the J.I so the existing assembly line would remain busy instead of stopping production while a completely new design was brought into production. Moreover, the many components already manufactured could be used in the updated design instead of having to be scrapped. Redesigned armor in the lightened J-type would reduce the weight by 100 kg. By using the improved wing structure developed for the Junkers CL.I and converting the rear fuselage from fabric to dural-covering the evolved design could be substantially lighter than the J.I. The camera and wireless equipment fixtures could be eliminated to save weight because these were rarely used on missions. To shorten the take-off run, the wing area was to be increased by increasing the span, although this would reduce maneuverability by further limiting the roll rate.

A modified Junkers J.I (J.749/18), fitted with a 260 hp Mercedes and dural-covered fuselage left the factory in early December 1918. This lighter, more powerful prototype was probably built as a result of the October staff conference. It was apparently the J.II development prototype although no documentation has been found to confirm this, and photographs taken of this aircraft on 14 and 22 January 1919 are the only proof of its existence.

(3) Military aircraft deliveries were stopped by the German government as of November 21, 1918, but emergency relief measures allowed production of aircraft to continue through January 1919 to provide work for the factory workers.

Left: Junkers J.I 102/17 was the third production machine. The Junkers J.I was very highly regarded by its crews for its extreme survivability in a very hostile environment, but was too heavy to carry the additional downward-firing machine guns eventually fitted to AEG and Albatros J-types, limiting its role to that originally envisioned by *Idflieg*; an infantry cooperation aircraft. Their extra weapons enabled the AEG and Albatros to perform ground-attack missions.

Above: Junkers J.I J.146/17 ready for another mission. It has late-style national insignia and two-color camouflage on its upper surfaces. It is fitted with an Axial propeller.

Below: Junkers J.I J.119/17 of *Flieger Abteilung (A)* 217 displays colorful markings for a Junkers J.I.

Above & Below: Junkers J.I J.849/17 tested the installation of a more powerful 260 hp Mercedes D.IVa engine. The additional power gave shorter take-off distance and improved rate of climb and maneuverability, but was not adopted for the J.I. With the Mercedes engine the nose contours were altered and no spinner was fitted. This aircraft remained at the Junkers factory for flight evaluation.

Above: Junkers J.I J.749/18 was a development aircraft with lightened airframe, revised tail surfaces, and 260 hp Mercedes D.IVa engine. This photo was taken on 14 February 1918 on the Dessau airfield. This aircraft was the prototype for what would have been called the Junkers J.II had it gone into production, but that designation does not appear in Junkers or *Idflieg* records.

Below: Here Junkers J.I J.749/18 is fitted with wings of increased span for comparison with the standard wings. This photo was taken on 22 February 1918.

Junkers J.I *Pour le Mérite* Observer

Above: Observer *Lt.* Wilhelm Paul Schreiber (at right) and his pilot, *Feldwebel* Ernst Schäfer, on the upper wing of their Junkers J.I J.134/17 while serving with *Flieger-Abteilung (A)* 221 in the Spring of 1918. They had flown together as a team since March 1917. On 15 May 1918 Schreiber shot down American Charles Biddle of the 103rd Aero Squadron while he and Schäfer were flying this aircraft (Biddle survived to become an ace). On May 30, 1918 Schäfer was hit in the head by ground fire and J.134/17 crashed between the trench lines, killing Schreiber as well. This crew had performed excellent work for many months and Schreiber had been nominated for the *Pour le Mérite*. Unusually, the Kaiser approved the award posthumously, a rare honor, and Schäfer was posthumously awarded the *Goldene Militär-Verdienstkreuz* (Military Service Cross in Gold, the highest award to an enlisted man). Before their fatal flight their trusty J.134/17 had accumulated 420 hits in previous missions. (Hanns-Gerd Rabe via Peter Kilduff)

At the time of his combat with Schreiber and Schäfer, Charles Biddle had scored two victories, and after his recovery from wounds from the combat with their J.I he went on to score five more, bringing his total to seven. In 1919 Biddle wrote *The Way of the Eagle*. Here is how Biddle described his encounter with the Junkers J.I 134/17 of Schreiber and Schäfer; he was flying Spad VII S.3149 of the 103rd Aero Squadron:

"...We were flying along inside our own lines, when I noticed a peculiar two-seater circling very low down between the trenches, he could not have been more than 600 meters up. I took him for an English infantry liaison machine (RE8 – ed.), which he very much resembled, but then noticed that he seemed to circle into the Boche lines with remarkable impunity considering his very low altitude, so decided to investigate. Sure enough there were the old black crosses on him showing plainly as he swung almost under me in making a turn over our lines... The only way that I knew that my friend was really a Hun was by his black crosses, for it was the first Boche machine of the kind that I had ever seen, and indeed I never heard of anyone that I know, running into one like it. He had a rounded body like some French machines, the tail was square and the lower wing much shorter than the upper, like many of the English two-seater observation planes... In addition to this it was the slowest bus you ever saw and I think I could go two miles to his one. All this leads me to believe that it was a new type of German armored plane which they call a "Junkers" and which I have read about in the aviation reports. They are built especially for infantry liaison work and are heavily armored about the fuselage to protect them from fire from the ground. In consequence of their great weight they cannot go very high and are extremely slow. This fellow must have been a squadron leader or something, for he had four big streamers attached to his wings, one on the top and another on the lower plane on each side. Perhaps, however, these may have merely been means of identification for the benefit of his own side... (that is correct – many *Infanterieflieger* airplanes carried such streamers – ed.]...He certainly got the best of me, and I don't feel at all vindictive about it, as it was a perfectly fair fight, but just the same it would give me more satisfaction to bring that boy down than five others..."

Biddle goes on to tell how he tried to get into the Junkers' blind spot by flying beneath it, but it was so slow he overshot and wound up too close to bring his gun to bear. The Junkers was "swerving from side to side to get me out from under him so that the machine gunner could shoot....[and] the observer, who proceeded to do the quickest and most accurate bit of shooting that I have yet run up against... I saw a black-helmeted head appear over the edge of the Hun machine and almost at the same instant he fired, as quickly as you could snap-shoot with a pistol..."

Biddle was hit in the knee, his Spad VII was hit in the motor, which stopped, and he crash-landed in no-man's land. He managed to make it to the British trenches and recovered in hospital where he wrote the above letter on May 25, 1918.

Junkers J.I J.134/17 of *Flieger-Abteilung (A)* 221 of pilot *Feldwebel* Ernst Schäfer and observer *Lt.* Wilhelm Paul Schreiber, Spring 1918.

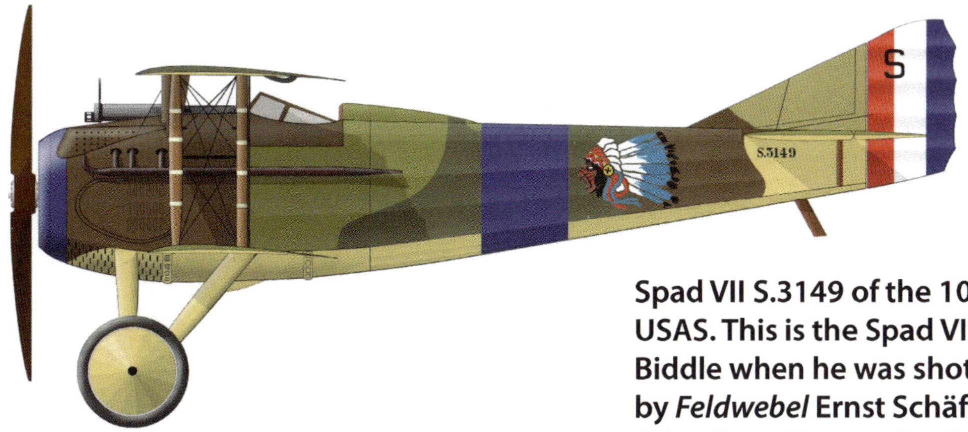

Spad VII S.3149 of the 103rd Aero Squadron, USAS. This is the Spad VII flown by Charles Biddle when he was shot down on 15 May 1918 by *Feldwebel* Ernst Schäfer and *Lt.* Wilhelm Paul Schreiber, *Flieger-Abteilung (A)* 221.

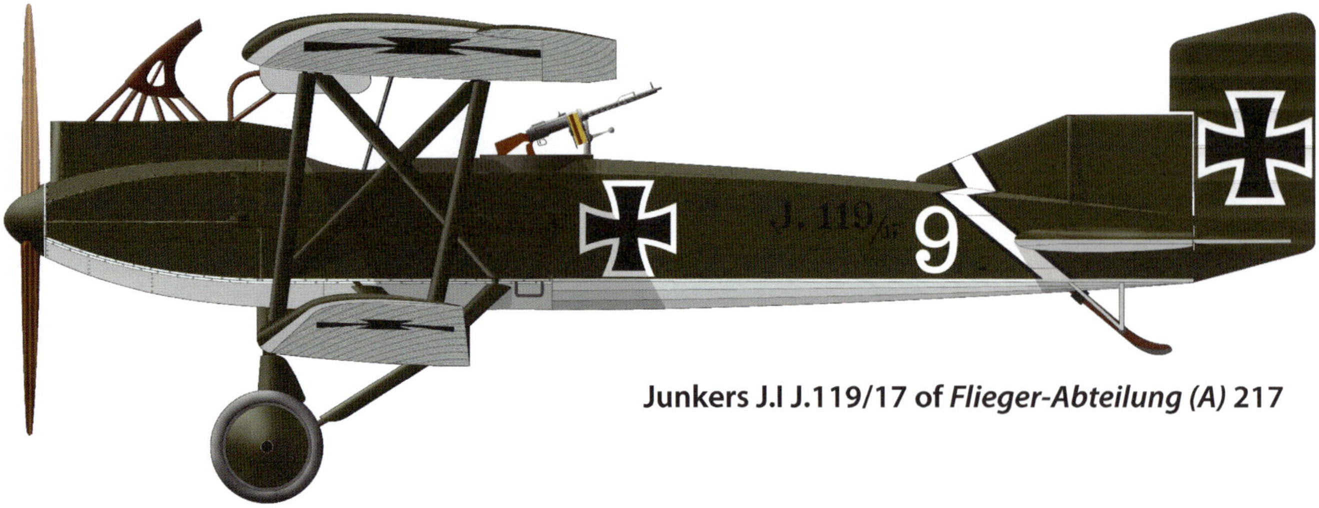

Junkers J.I J.119/17 of *Flieger-Abteilung (A)* 217

Junkers J.I J.802/17

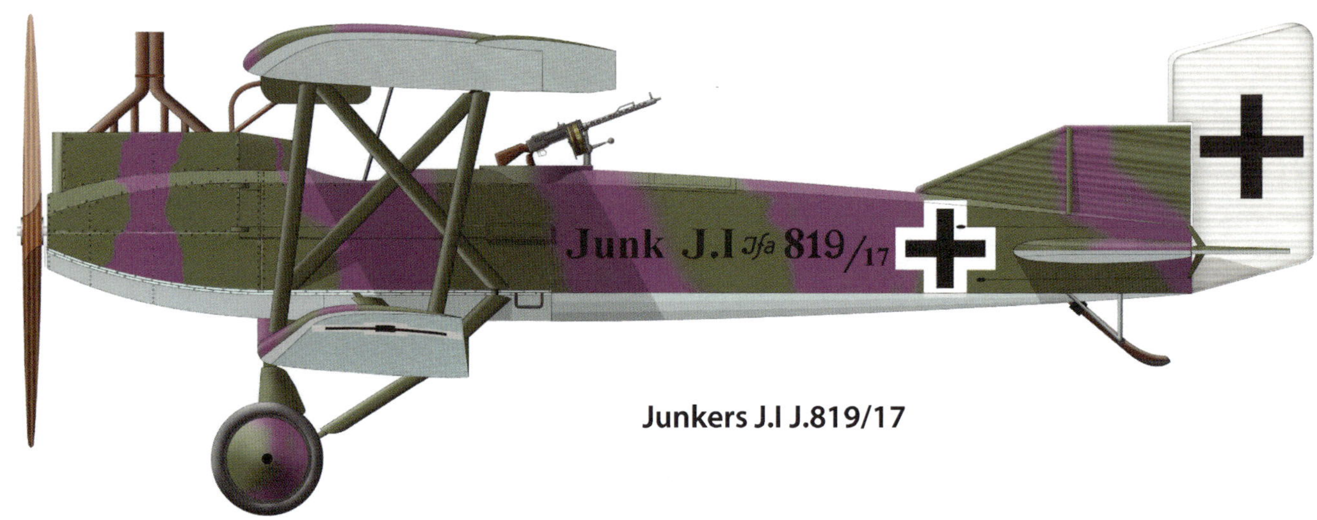

Junkers J.I J.819/17

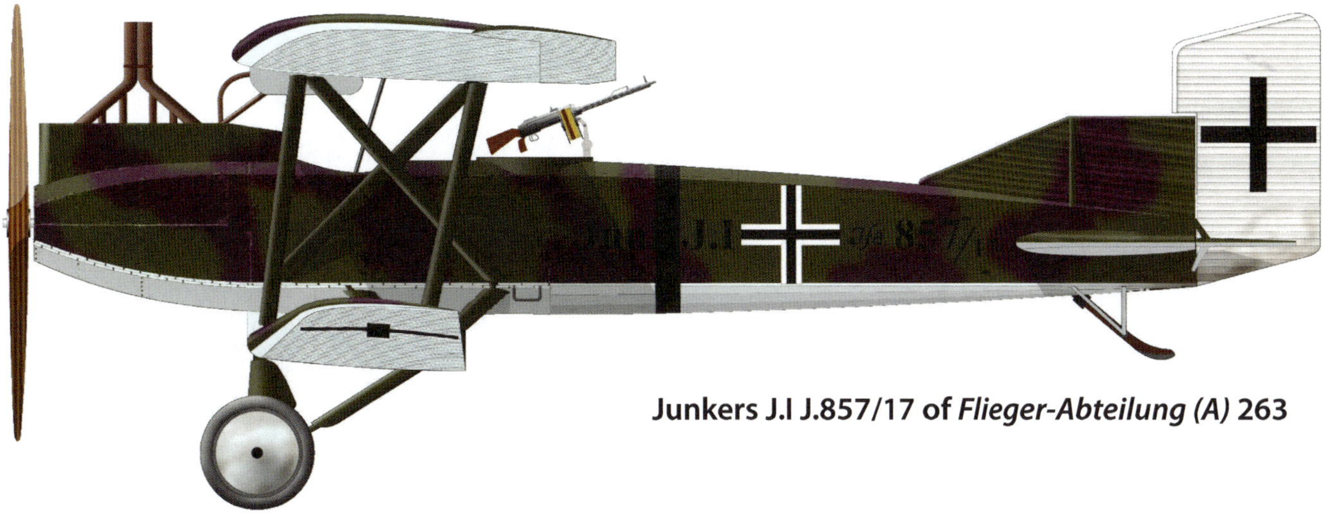

Junkers J.I J.857/17 of *Flieger-Abteilung (A)* 263

Junkers J.I J.586/18, now displayed in Canada

Junkers J.I J.596/18

Above: Junkers J.I J.586/18 showing the printed camouflage fabric covering the rear fuselage. The British wing in the background indicates the photo was taken postwar. This is the aircraft now displayed in Canada.

Above: This abandoned Junkers J.I photographed postwar may be J.884/17; see additional photo at top on page 83. The metal structure of the rear fuselage is shown to advantage.

Above: Derelict Junkers J.I J.884/17 photographed on the Bickendorf aerodrome near Cologne in early 1919. When nearly all other contemporary airplanes were made of wood, the all-metal Junkers J.I was a marvel, and many postwar photographs were taken by impressed Allied soldiers. The aircraft were also souvenired, resulting in the aircraft becoming increasingly abused. Additional photos are below right and at the bottom of page 83.

Above: The nose armor of the Junkers J.I is shown in this postwar photograph.

Above: Another view of Junkers J.I J.884/17 photographed on the Bickendorf aerodrome near Cologne in early 1919.

Above: Derelict Junkers J.I fuselages postwar, probably at Romorantin.

Left: Junkers J.I fuselages being prepared for shipment back to the USA after the war. The photo was taken at Romorantin, and may be the two machines seen above.

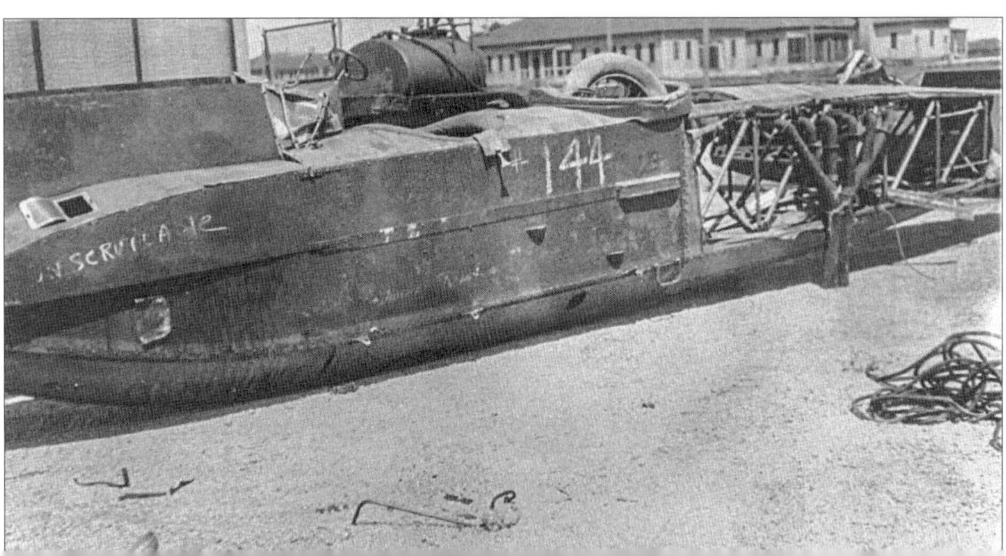

Left: Derelict Junkers J.I fuselage postwar shows the rear fuselage structure.

Sketches of Junkers J.I Details from the Official Air Ministry Technical Report, 1920

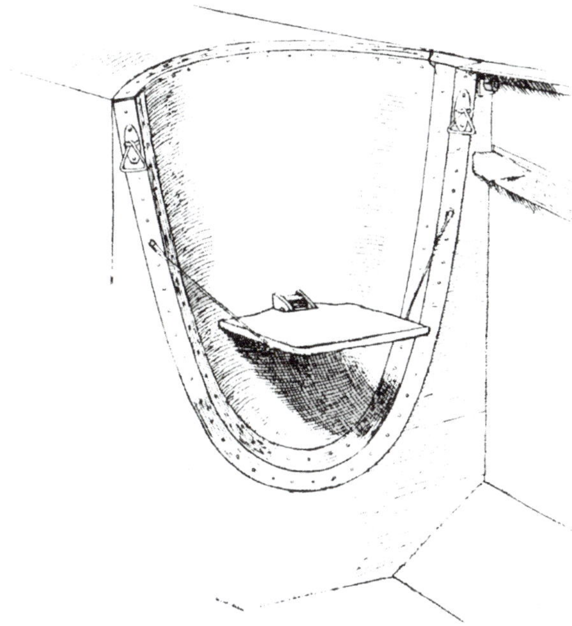

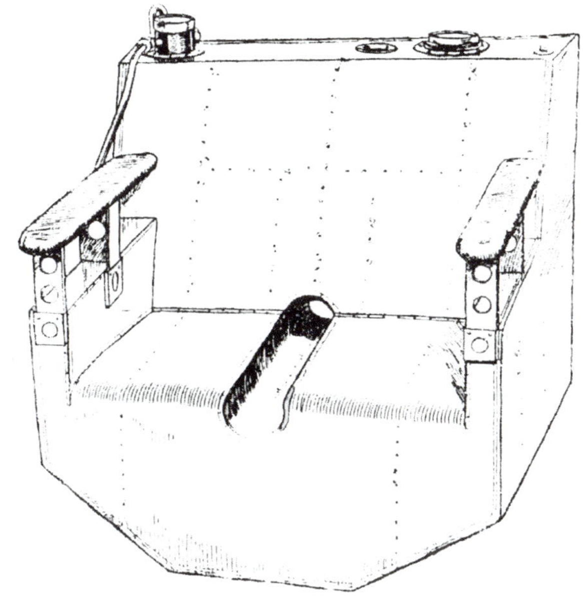

Above: Looking aft, the folding gunner's seat attached to the rear of the armored section of the fuselage.

Above: The pilot's seat was also the main fuel tank. Made of brass and having 26-gallon capacity, it had a padded seat and back. The hole was for the elevator control rod.

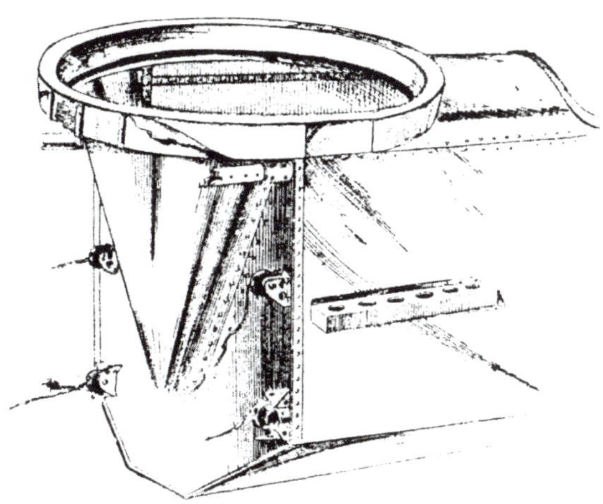

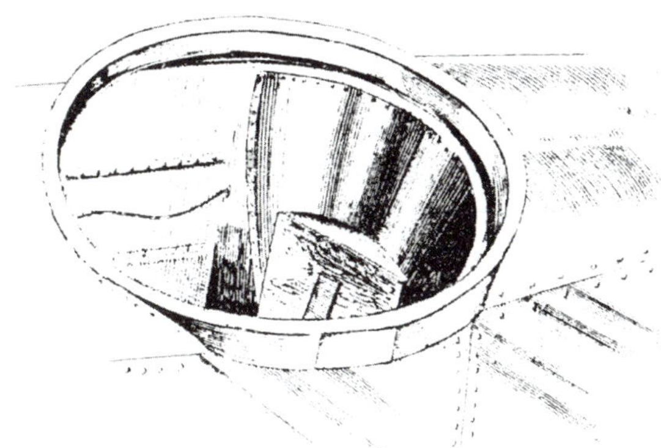

Above: Looking aft, the gunner's cockpit.
Left: Rear side view of the gunner's cockpit armor.

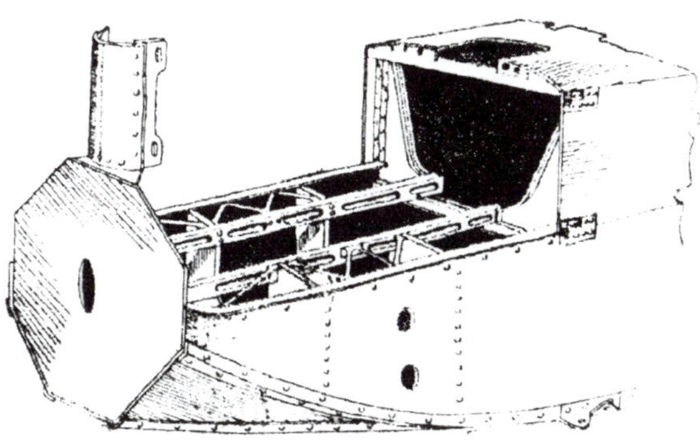

Above & Right: Details of the armored engine compartment and access panels.

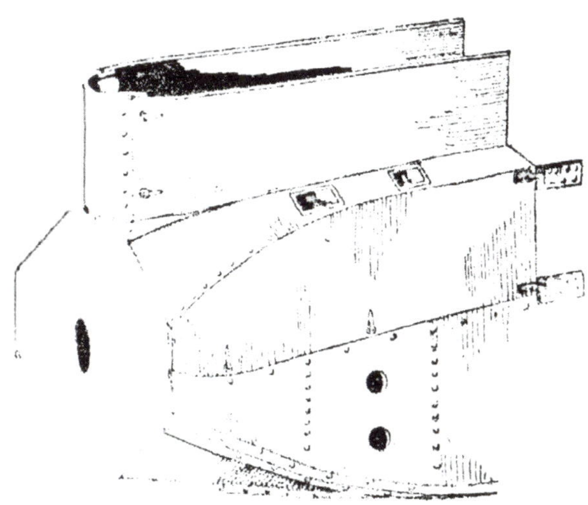

Other German Armored Aircraft

Above & Below: The AEG DJ.I triplane was envisioned as a low-altitude fighter armored to withstand ground-fire as it intercepted enemy ground-attack aircraft at low altitude, a role similar to that foreseen for the Sopwith Salamander. It was powered by a 195 hp Benz Bz.IIIb V-8 engine, which did not reach production in time to power any warplane over the front.

AEG DJ.I Triplane

Exactly when the idea of an armored, single-seat fighter to oppose Allied ground-attack aircraft was proposed is not known. Surely the concept was another inspiration of *Oberst-Leutnant* Wilhelm Siegert, the commander of *Idflieg*, who throughout the war remained a steadfast proponent of specialized aircraft to fulfill specific tactical assignments. He had no use for general-purpose aircraft. It was reported on 21 October 1917 that AEG was designing an armored fighter, known internally as the AEG PE (*Panzer-Einsitzer*) and may have been later designated DJ.I or DA.I. (In late 1918, two aircraft classifications appeared in various records, both signifying the armored single-seat fighter. They were DJ = D-*Infantrie* and DA = D-Angriff (*attack*). It is not known if these classifications were officially adopted.) The engine, fuel, and pilot were protected by a steel shell forming an integral part of the airframe. The rear fuselage was of aluminum. The triplane wings had AEG's customary steel-tube spars and wooden ribs.

The preliminary trials of the AEG PE triplane, powered by a 195 hp Benz Bz.IIIb V-8 engine, were expected to be concluded in December 1917. From a report dated March 1918, we learn that the AEG *Panzer-Einsitzer* had reached 4000 meters (13,123 ft.) in 48 minutes. The test pilot reported the triplane as easy to fly but not particularly fast. Work on the AEG DJ.I triplane was stopped in May 1918 because of "poor flight characteristics."

Above: The AEG DJ.I biplane was developed from the earlier DJ.I triplane to improve its performance and flight characteristics. It was powered by the same experimental 195 hp Benz Bz.IIIb V-8 engine, which likely delayed its development. The biplane use the same armored fuselage as the triplane coupled to thick biplane wings unlike those on any other AEG design of the war.

AEG DJ.I Biplane

Disappointed by the mediocre flight characteristics and low speed of the AEG DJ.I triplane, AEG engineers brought out a biplane version, the AEG DJ.I fighter, also powered by the 195 hp Benz Bz.IIIb V-8 engine. Sharing the triplane's armored fuselage design, the DJ.I had a wireless wing cellule to minimize damage from ground fire. The construction began in March and was completed in May 1918, but according to *Idflieg* it was not until August that the flight tests began. The reason for the delay is not known but may have been caused by unavailability of the experimental Benz V-8 engine. Converting to a biplane wing cellule improved the aircraft's speed and climb rate. In a 1920 aircraft inventory compiled by *Luftfriko* there were two AEG DJ.I aircraft listed as "property of the factory." Whether this referred to the DJ.I triplane and biplane or the second DJ.I fighter described below has not been determined.

A second AEG DJ.I powered by a 245 hp Maybach Mb.IVa engine was under construction from May to July 1918. In September we learn that the Maybach engine was to be replaced by a new one, pointing to the probability that the second DJ.I had performed some flight trials.

A third AEG DJ.I was reported under construction in September 1918, but there is no record of completion. That three DJ.I prototype fighters existed was consistent with the *Idflieg* policy of ordering one prototype for flight tests, one for static-load testing, and one held in reserve.

Above & Below: The AEG DJ.I biplane was well streamlined and was designed with thicker wings to enable wireless bracing, reducing drag. The 195 hp Benz Bz.IIIb V-8 engine drove a four-bladed propeller.

Above: The AEG G.IVk was a modification of the standard G.IV bomber with armored engine nacelles and forward fuselage crew positions plus a 20mm Becker cannon for the forward and aft gunners for anti-tank duties. Each gunner also had a flexible machine gun to defend against fighter attacks.

AEG G.IVk

Another German armored aircraft type designed for ground-attack was the AEG G.IVk, which was developed from the standard AEG G.IV bomber. AEG G.IV G.1096/16 was armed with a 20mm Becker cannon mounted in an experimental rotating chin turret to investigate the feasibility of attacking tanks and other ground targets from the air. The field of fire was 50 degrees vertical and 90 degrees horizontal. *Idflieg* reported that the aircraft was slated for combat evaluation in December 1917, but further information is lacking. However, ground and air trials demonstrated that the visibility, field of fire, and target acquisition were satisfactory. The cannon installation was reported as "frontline ready", leading to the development of the specialized AEG G.IVk ground-attack aircraft.

As a result of the progress made with G.1096/16, in March 1918 *Idflieg* ordered five AEG G.IVk grount-attack aircraft for service trials. Numbered G.500/18 to G.504/18, they were armed with two 20mm Becker cannons, one mounted in the lower nose turret and a second in a rear turret that allowed for firing over the fuselage or downward through a 'Gotha-type' tunnel. For defense against fighters, two Parabellum machine guns were mounted on LVG gun rings (copies of the Scarff) in the nose and rear turrets. Like the rear cannon, the rear Parabellum could also fire downward through the fuselage tunnel. Four 50 kg PuW bombs could be carried under the fuselage. Four vertical magazines in the front cockpit and three in the rear held 12.5 kg PuW fragmentation bombs. Chrome-nickel steel armor plating 5 mm thick covered the floor and sides of the fuselage, extending from the nose around the rear gunner's cockpit. The engine nacelles were armored as well.

Flight trials of the prototype, G.500/18, in February 1918 demonstrated good results but the biplane tail and armor plating required minor changes. The cannon firing trials performed in April 1918 were satisfactory. Though dispatched to the front in October–November 1918, it is not known if the G.IVk saw operational service. As part of the Armistice stipulations, the Germans turned over the four production AEG G.IVk ground-attack aircraft (G.501/18–G.504/18) to the British in January 1919.

This Page: Four production AEG G.IVk ground-attack aircraft, G.501/18–G.504/18, were turned over to the British as part of the armistice terms; G.503/18 appears to the the aircraft shown here.

Above: The prototype AEG G.IVk was G.500/18 shown here on the AEG factory airfield at Henningsdorf. The four production aircraft had a different tail design; see below for comparision. G.500/18 wears an interesting camouflage of two-color sprayed camouflage on the fuselage, dark engine nacelles, and printed camouflage fabric on the fins and rudders. The 20mm Becker cannon in the nose mounting is prominent. This aircraft was dispatched to the front for operational use in October–November 1918, but it is not known if it was used in combat.

Above: The AEG G.IVk G.503/18 was one of the four production G.IVk ground-attack aircraft turned over to the British according to the terms of the armistice. The vertical tail surfaces of the four production machines differed from the prototype. The aft gunner could fire his 20mm Becker cannon conventionally to the sides or down through the fuselage in a Gotha-like tunnel, and could do the same with his machine gun. The forward gunner had a forward-firing 20mm Becker cannon for anti-tank duties and a flexible machine gun to defend against fighter attack. No conventional bomb bay was fitted but the AEG G.IVk had four vertical tubes about two feet long mounted in the front cockpit and three in the rear cockpit. These tubes were perhaps intended for launching flares or special anti-tank bombs that *Idflieg* had developed.

Above: The nose turret of the AEG G.IVk showing the 20mm Becker cannon. Armored doors to close the mounting are at the sides of the opening for the cannon.

Above: The rear gunner of the AEG G.IVk in his turret handling his 20mm Becker cannon. Although no bomb bay was fitted and supposedly no bomb racks, bombs can be seen underneath his position. He could fire both his cannon and machine gun to the sides and downward through a Gotha-style tunnel through the fuselage.

Above: The nose turret of the AEG G.IVk with the 20mm Becker cannon seen from the inside. Although the cannon pointed forward and down for ground-attack, the gunner's machine gun was in a conventional gun ring to defend against fighters.

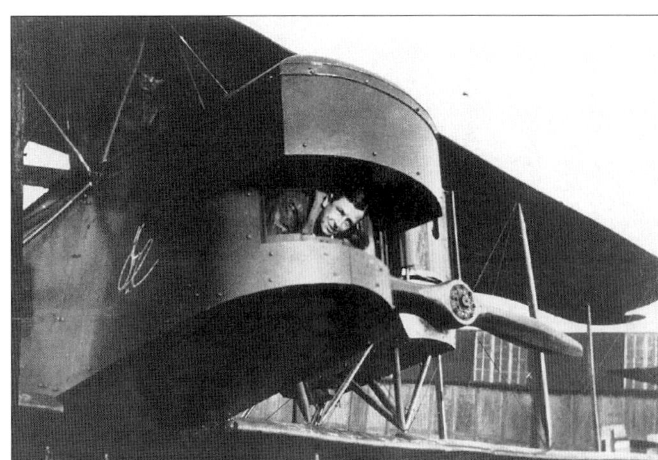

Above: An Australian soldier pokes his head our of the nose turret of the AEG G.IVk; another stands in the cockpit. The armor plate was difficult to form, so all armor on the aircraft was either a flat plate or a simple curve. Compound curves were simply too difficult to form at the time.

AEG G.IVk G.500/18

AEG G.IVk G.503/18

Above: The massive Ago S.I was the only S-type aircraft completed before the armistice. Heavily armed and armored, it was the first of a new class of heavy ground-attack aircraft armed with a 20mm Becker cannon to destroy tanks. It was powered by the 300 hp Basse & Selve BuS.IVa engine.

Ago S.I

The Ago S.I was the only S-type (*Schlachtflugzeug* for battleplane, or heavy ground-attack aircraft) to be completed during the war. The primary mission of the S-type, which evolved from the J-type, was ground attack, with an emphasis on destroying tanks. The *Idflieg* aircraft department and weapons command coordinated with Ago engineer Rudolf Blaser in designing the S.I. Powered by a 300 hp Basse & Selve BuS.IVa engine, the S.I had a gunner's location beneath the fuselage to provide the maximum field of fire against ground targets for the flexible 20mm Becker cannon (and/or twin machine guns). The pilot, gunner, engine, and fuel tank were all protected by armor varying between 3mm and 7mm in thickness. The pilot's head was protected by a hinged, armored hood. In June 1918 the S.I airframe was inspected by *Idflieg* engineers and the final armor configuration determined. Ago received German patent No. 329,223 on 11 July 1918 for the S.I configuration. Flight tests, slated to begin in October 1918, were purportedly performed by Ago test pilot Lorinsen; this is contradicted by the Inter-Allied Commission, who reported that the two Ago S.I aircraft never flew. Unfortunately, dimensions and technical details of the unique S.I are not available.

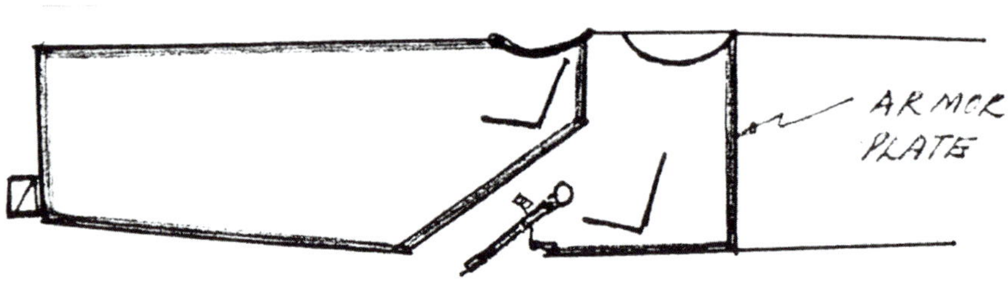

Ago S.I Armor & Cannon: Patent 329,223, 11 July 1918

Left: This patent drawing for the Ago S.I shows the configuration for the armor and the location of the 20mm Becker cannon. The gunner also had a flexible machine gun to defend against fighter attack.

Above: The Ago S.I was the only S-type aircraft completed before the armistice. To reduce control forces in this massive aircraft, the control surfaces featured horn balances, improving maneuverability. The undercarriage was designed to provide an unobstructed field of fire for the flexible 20mm Becker cannon mounted to fire forward and down.

Above: Albatros J.I 09.03 was the last of three purchased by Austria-Hungary. Formerly J.730/17, 09.03 was the only J.I used operationally by the *Luftfahrtruppe;* it served with *Flik* 69/S between July and September 1918.

Fokker V37

Derived from the earlier Fokker V27, and like it powered by a 195 hp Benz Bz.IIIb liquid-cooled V-8 engine, the Fokker V37 was an armored fighter designed for low-level fighting and ground attack. The V27 was itself clearly inspired by Fokker's earlier E.V/D.VIII fighter.

Clearly the radiator had to be protected from ground fire to make an armored airplane effective, and the V37's solution was clever. The radiator was mounted behind the propeller and protected by a disk of armor. The propeller was given a large spinner for streamlining, and air was forced through the radiator by six fan blades mounted in the plane of the propeller. The entire forward fuselage, including the engine and cockpit, was protected by steel armor plate 2.5mm thick. The armor behind the pilot even extended up behind his head and was faired into a headrest.

Had the V37 design progressed, it would certainly have been fitted with the standard fighter armament of two synchronized machine guns. However, the V37 arrived too late to be evaluated, let alone produced in quantity, so its performance and flying qualities are not known.

Above: The Fokker V37 armored fighter prototype combined the configuration of the earlier Fokker E.V/D.VIII with a distinctive propeller spinner and a 195 hp Benz Bz.IIIb V-8 water-cooled engine.

Below: The entire forward fuselage of the Fokker V37 was armored with 2.5mm steel to protect the pilot, engine, and radiator from rifle-caliber ground fire.

Above & Below: These views of the Fokker V37 armored fighter prototype emphasize its streamlined form and great resemblence to the earlier Fokker E.V/D.VIII.

Right: This view of the Fokker V37 armored fighter prototype without wing or fabric covering clearly shows its armored front fuselage and propeller disk and its welded steel tube fuselage contruction. The cooling fan blades in the plane of the propeller are visible. The back of the pilot's head was even protected by a tab of armor extending up from the back armor. The cockpit was very close to the engine to minimize the amount of armor that was needed to protect them.

Armored Aircraft in Retrospect

J-Type Summary

The J-type *Frontbeststand* inventory on page 4 provides useful insight into German armored aircraft. The AEG J.I appeared at the front first, undoubtedly due to the ease with which it could be put into production compared to the complex, all-metal Junkers that demanded new manufacturing technology. AEG continuously developed its J-types, resulting in gradual evolution of the J.I into the final production AEG J.II. The more agile AEG J.II started replacing the AEG J.I at the front in Spring 1918 and went on to become the most numerous J-type in combat. Although not as robust as the all-metal Junkers J.I, it was more heavily armed for offensive use and easier to produce. If the Junkers excelled at the original infantry mission, the AEG J.II was the best ground-attack machine.

The Junkers J.I was the second J-type to reach the front. Its numbers grew slowly due to the difficulty of producing this revolutionary aircraft. Sustained effort gradually improved the Junkers J.I production rate. Yet, despite the great operational success of the tough Junkers J.I, it was out-numbered at the front by the more easily produced AEG J-types even though these were not as resistant to ground fire. Interestingly, despite its deserved reputation as the best J-type due to its survivability, the Junkers was the only J-type that did not adopt oblique-firing machine guns for trench strafing. A look at the specification table gives an important clue why; the Junkers was by far the heaviest aircraft of an under-powered group. Aircrews already regarded it as a 'moving van', and the additional weight of more guns and ammunition likely could not be tolerated. However, despite its weight the Junkers was the fastest of the J-types due to being aerodynamically cleaner than the AEG and Albatros designs, largely due to its lack of bracing wires.

Significantly, the powerful Albatros firm produced the fewest J-types; moreover, they were the last to reach the front. Given the huge production capacity of Albatros, this is a strong indication that the Junkers and AEG J-types were superior to the Albatros J-types. Certainly the Albatros J.I, with its exposed radiator and no engine armor, was the J-type most vulnerable to ground fire, and thus the least successful. Although the Albatros J.II rectified those weaknesses, the wood structure of the Albatros J-types made them more vulnerable to ground fire than the metal Junkers and AEG designs; therefore they were produced in the smallest numbers despite Albatros's substantial production capacity.

J-Type Technology

Introduction of a mass-produced, all-metal aircraft over the Western Front represented a major technological breakthrough. Although it was an unlovely design of modest speed, many aircrews owed their survival to the tough Junkers J.I. Although historically overshadowed by glamorous fighter aces and their colorful aircraft, the Junkers J.I previewed future aircraft structures.

Despite its installation in all operational J-types, the Benz B.IV/IVa of 200/220 hp was clearly insufficient for these heavy aircraft. In comparison, the Liberty Plane was hundred kilograms lighter, yet had a 400 hp Liberty engine. The American armored LUSAGH-21 prototype with 400 hp Liberty, developed in response to the J-types' success, was rejected due to inadequate performance.

Thus the J-types were significantly underpowered even by the modest standards of the time, resulting in long take-off runs, thereby reducing the number of airfields that could be used. Their low power also resulted in low speed, which, together with their higher stall speed due to their heavy armor, provided a relatively small maneuvering speed margin above the stall compared to their unarmored contemporaries. This made them less forgiving to fly and reduced their safety margin during normal flight operations. Being underpowered also reduced their agility and made them more vulnerable to fighters.

Aware of aircrew criticism, in April 1918 *Idflieg* suggested use of the more powerful 260 hp Mercedes D.IVa engine in J-type aircraft to improve their ability to operate from small, unimproved fields. Because the Mercedes D.IVa engine was not an over-compressed, high-altitude design, it was suitable for the low-flying J-types, and the AEG J.III prototype, designed in September, used this engine, as did Junkers J.I J.749/18. Additionally, both the AEG J.III and Junkers J.I J.749/18, the development prototype for the Junkers J.II, had redesigned armor plate that saved 100 kg (220 lb.) of weight. The combination of more power and less weight would have made these aircraft safer and more pleasant to fly, enabled the use of shorter fields for take-off, and improved their ability to defend themselves against fighter attack.

Use of the 200 hp Benz for these heavy aircraft begs the question why a more powerful engine was

Above: Survivability is the benefit of armored aircraft.

not used from the beginning. While *Idflieg's* theory was that J-type performance did not matter due to their armor, this was surely a rationalization forced on *Idflieg* by Germany's desperate resource situation. No military pilot has *ever* flown an aircraft in combat that was too fast! The distant blockade by the Royal Navy gradually starved the German economy. This forced *Idflieg* to economize in many areas, and the engines installed in the J-types were undoubtedly one of them. Despite the terrible battles on the Western Front and elsewhere, the reality is the Royal Navy won the war with its distant blockade that gradually but inexorably starved the German war machine until final Allied victory.

Armored Aircraft

The separation of combat roles initially envisioned between J-type and CL-type aircraft started to break down with combat. As experience was gained, the armored J-types began to be used more frequently for ground attack, the role originally assigned to the *Schlachtflugzeuge*, and most J-types acquired oblique-firing guns for ground attack. In particular, the armored aircraft began to be viewed more and more as anti-tank aircraft. The common image of massive German Panzer thrusts smashing through Allied lines and revolutionizing warfare came from bitter experience in a later war. In WWI it was the British who invented the tank to break the stalemate of trench warfare and the Germans who lacked a tank force able to counter this threat. Armored aircraft for tank-busting was an idea whose time had come in 1918 Germany, who proved unable to field a powerful tank force themselves, and a number of J-types were fitted with 20mm Becker cannon for tank-busting.

Furthermore, the armored S-type was developed from the J-type specifically for close air support, with an emphasis on destroying Allied tanks. The Ago S.I arrived too late to reach production, but with its flexible, downward-firing 20mm Becker cannon in an armored fuselage, it appears to be well-designed for anti-tank duties.

Larger armored aircraft were also developed for tank destroying. The twin-engine AEG G.IVk actually reached the front, though perhaps not quite in time to see combat. With engine nacelles and crew positions both armored like the J-types, plus flexible, downward-firing 20mm Becker cannon in gunners' positions fore and aft, it foreshadowed the modern attack helicopter and may have been similarly successful at destroying tanks had it reached combat in numbers. Armored, low-altitude fighter designs remained prototypes.

The Germans fielded both armored J-types and un-armored CL-types for ground-attack, while the Allies used the more numerous Sopwith Camel and similar un-armored fighters for close air support. With the exception of a dozen French Salmson 4 aircraft, the Allies used no purpose-built armored aircraft operationally, although armored prototypes like the Sopwith Salamander were evaluated.

Ever since WWI there has been a debate about the relative merit of specially-designed, armored aircraft and more general-purpose aircraft in ground attack. The specialized, armored airplane has generally lost that debate in favor of more numerous, faster, and versatile aircraft. However, in WW2 the armored IL-2 *Stormovik* was critical to the Soviet army, and the armored A-10 was very successful in both Gulf Wars. So the specialized, armored aircraft has generally been the more effective ground-attack aircraft. It is only with the availability of precision-guided munitions that general-purpose aircraft have become almost as effective as specialized, armored aircraft in ground attack. As debate and development continue, we remember that the successful German J-types were the first generation of armored, close air-support aircraft to see combat.

Above: This view of the inside of the Zeppelin hangar in Trier shows AEG J.II 353/18 tactical '7' with its wings still on. A USAS Salmson 2 A2 is at lower left with a Rumpler C.IV in the center, an LVG C.VI behind the AEG J.II, and Fokker D.VII fighters among the aircraft that can be identified. See page 31 for a closeup of J.II 353/18 without its wings.

Bibliography

Books

Gray, Peter, and Thetford, Owen, *German Aircraft of the First World War*, second revised edition, New York: Doubleday & Company, Inc., 1970.

Grosz, Peter M., *Junkers J.I*, Berkhamsted, Albatros Publications, 1993.

Lamberton, W.M., *Reconnaissance & Bomber Aircraft of the 1914–1918 War*, Letchworth, England, Harleyford Publications, 1962.

Articles

Grosz, Peter M., "An Armored AEG" *Windsock* Vol.11 Nr.6, Nov/Dec 1995.

Grosz, Peter M., "The 2 CM Becker Aircraft Cannon – Development and Use" *Over the Front* Vol.7 No.2, Summer 1992.

Grosz, Peter M., "Rare Birds: AEG J.Ia & J.II" *Over the Front* Vol.17 No.4, Winter 2002.

Herris, Jack, "Rare Birds: Albatros J.II" *Over the Front* Vol.23 No.3, Autumn 2008.

Schmeelke, Michael, "Rare Birds: Albatros J.I *Infanterieflugzeug*", *Over the Front* Vol.21 No.1, Spring 2006.

Rabe, Hans-Gerd, "Over the Lines in an AEG J.I," *Cross & Cockade* Vol.10 No.1, Spring 1969.

Rabe, Hans-Gerd, "A Forgotten *Pour le Mérite* Flier: *Leutnant* Wilhelm Paul Schreiber," *Cross & Cockade* Vol.15 No.2, Summer 1974.

Internet

http://en.wikipedia.org/wiki/Duralumin

http://en.wikipedia.org/wiki/Deutsche_Luft-Reederei

Above: This view of the war trophies exhibition in the Hamilton Armories during November 1919 shows Junkers J.I 586/18 at far right, AEG G.IV 574/18 in the background, and a Fokker D.VII at far left. The fuselage of Barker's Sopwith Snipe E8102 in which he won the Victoria Cross is in the center background amongst other captured German equipment.

Below: Unrestored Junkers J.I 586/18 in the Canada Aviation and Space Museum. It is the only Junkers J.I in existence.

Above: Another view of the inside of the Zeppelin hangar in Trier shows AEG J.II 353/18 tactical '7' with its wings on.

Above: An Albatros J-type undergoing load testing. It is probably a J.II given the late style insignia.

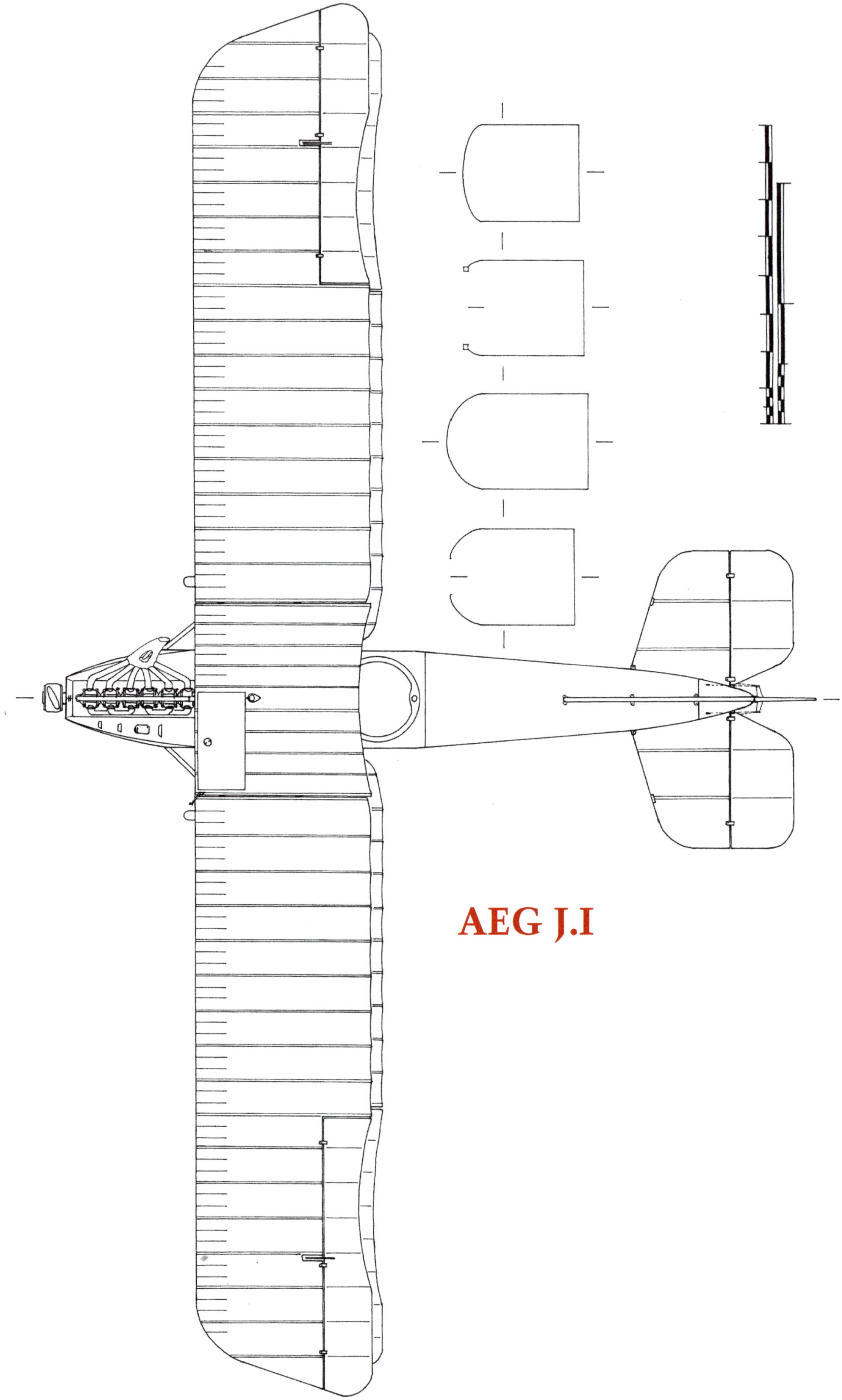

AEG J.I

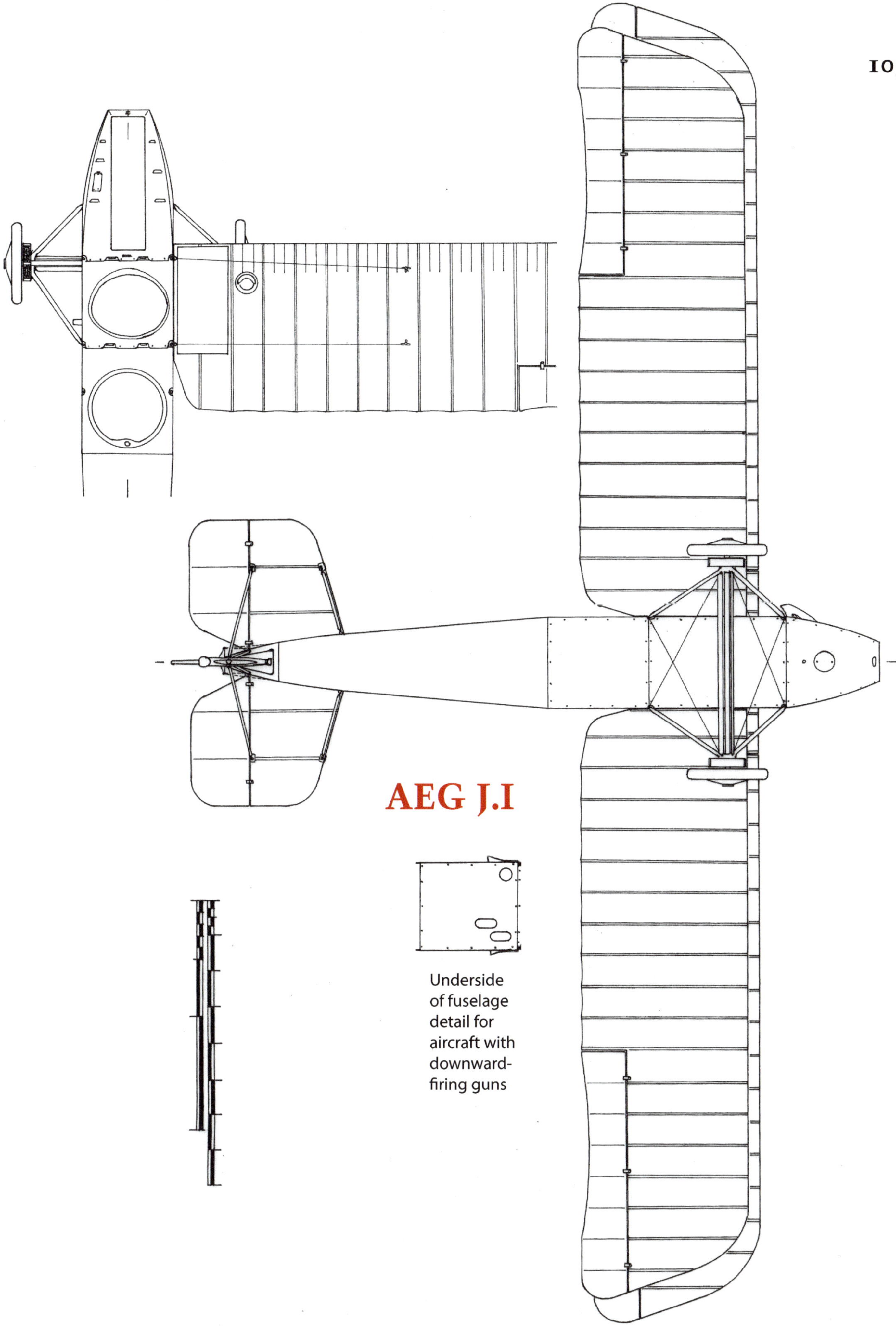

AEG J.I

Underside of fuselage detail for aircraft with downward-firing guns

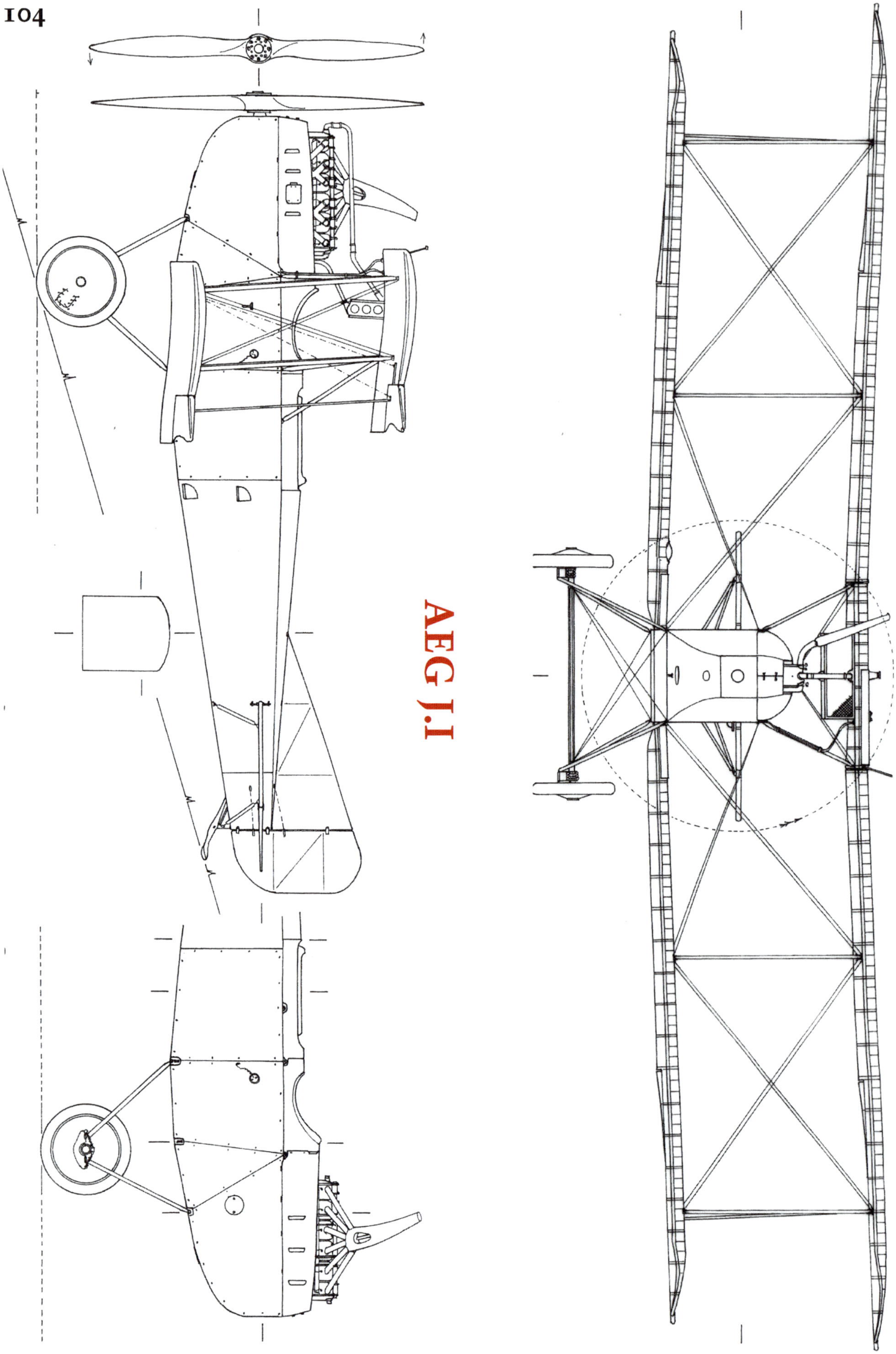

AEG J.I

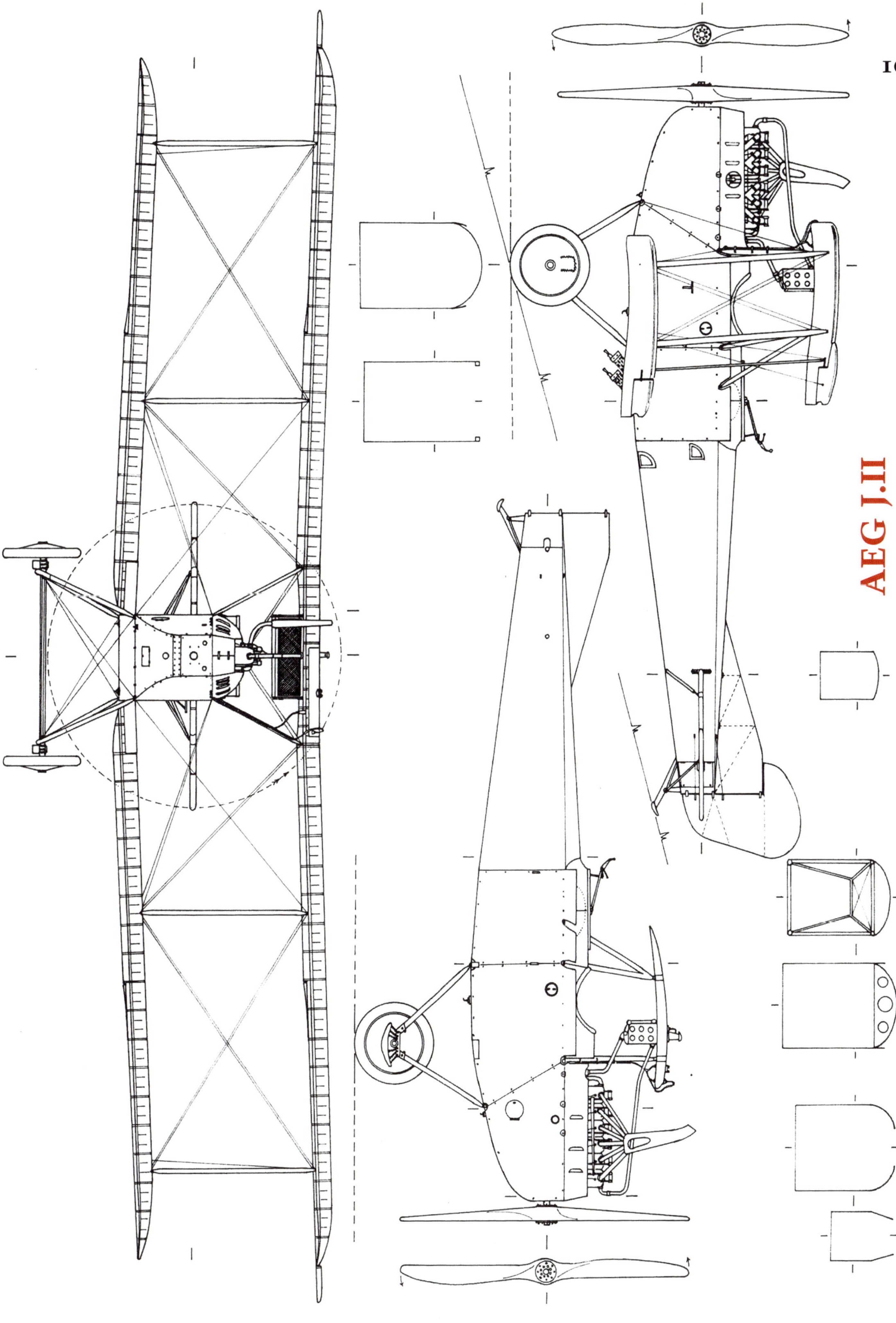

AEG J.II

AEG J.II

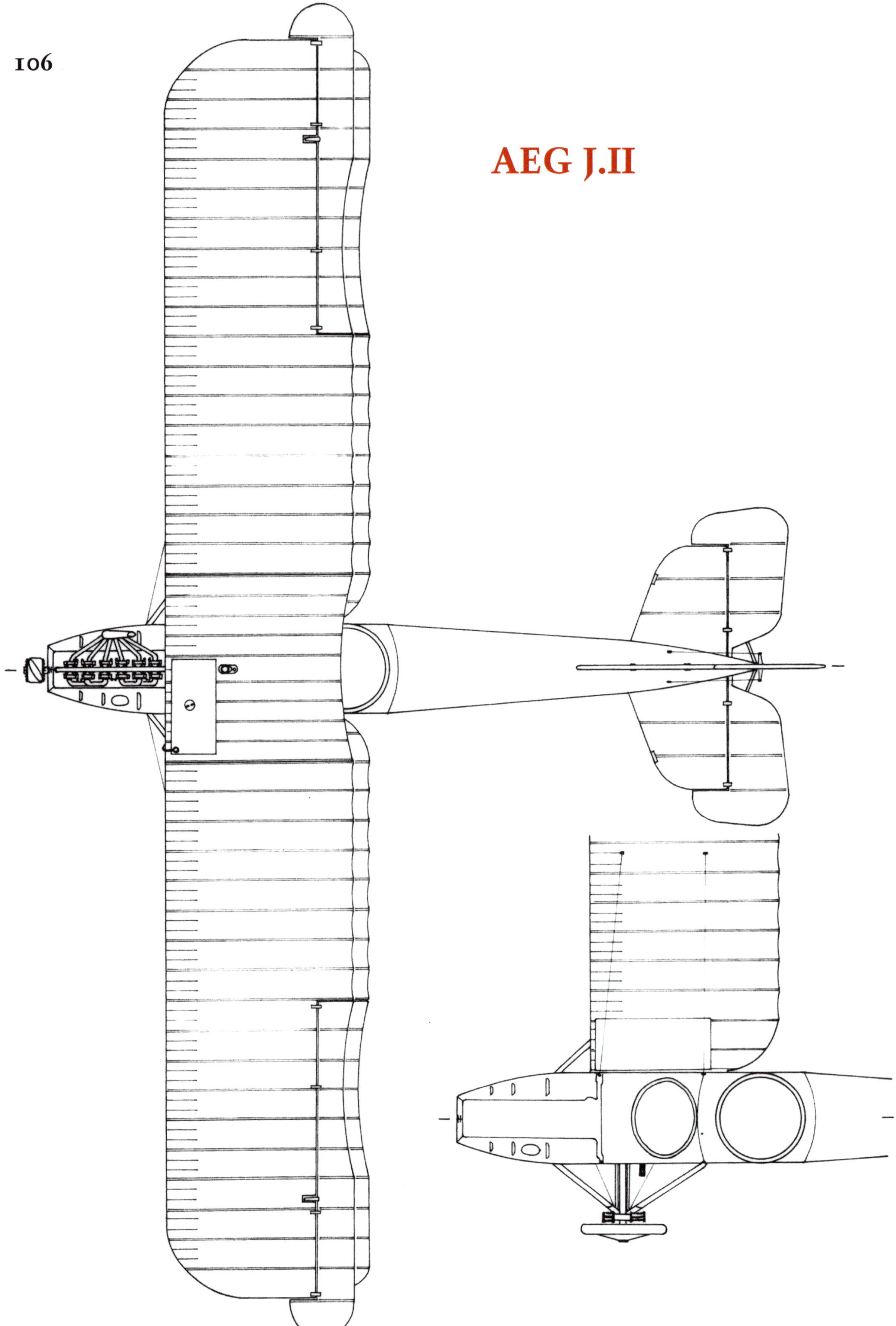

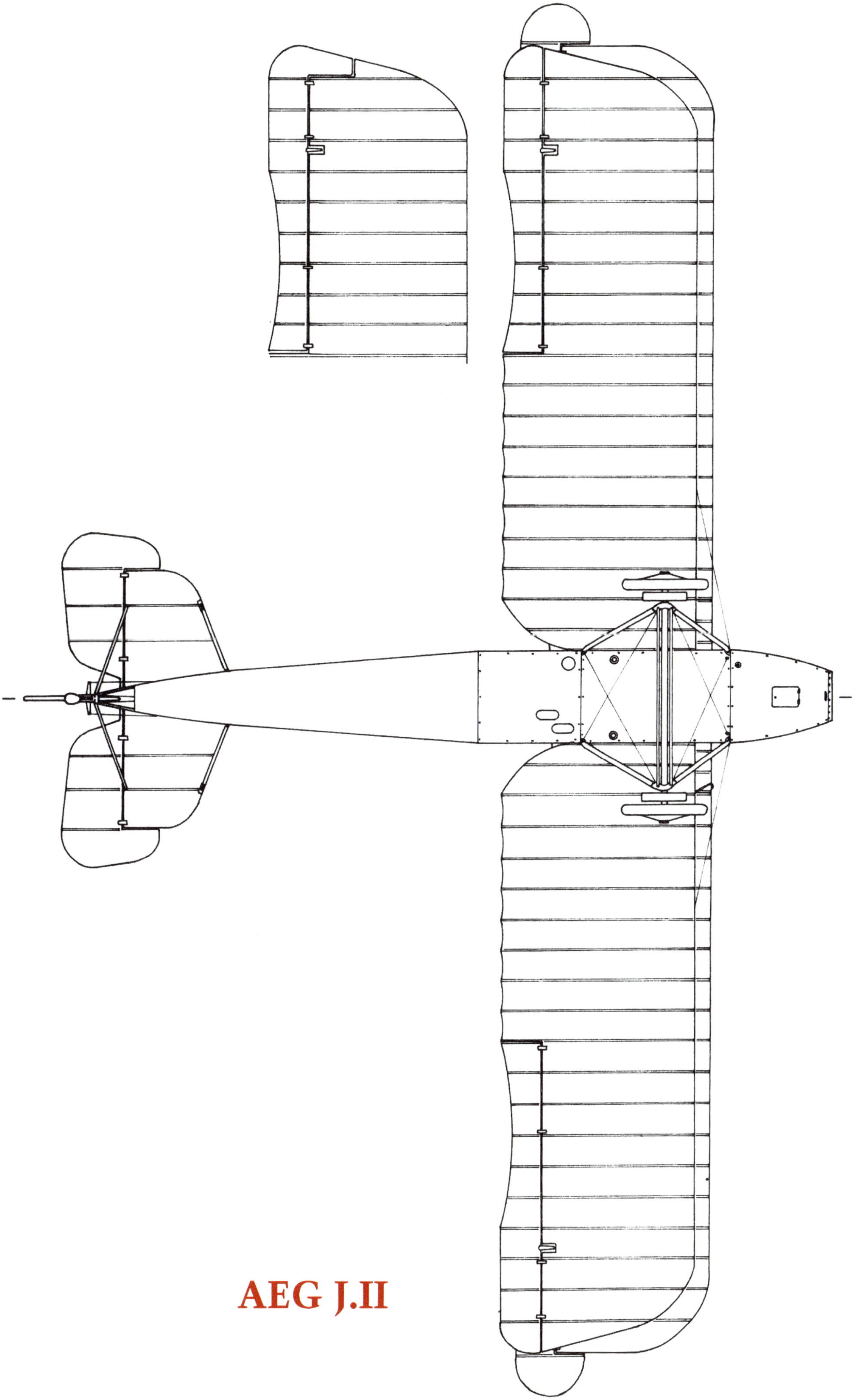

AEG J.II

Albatros J.I

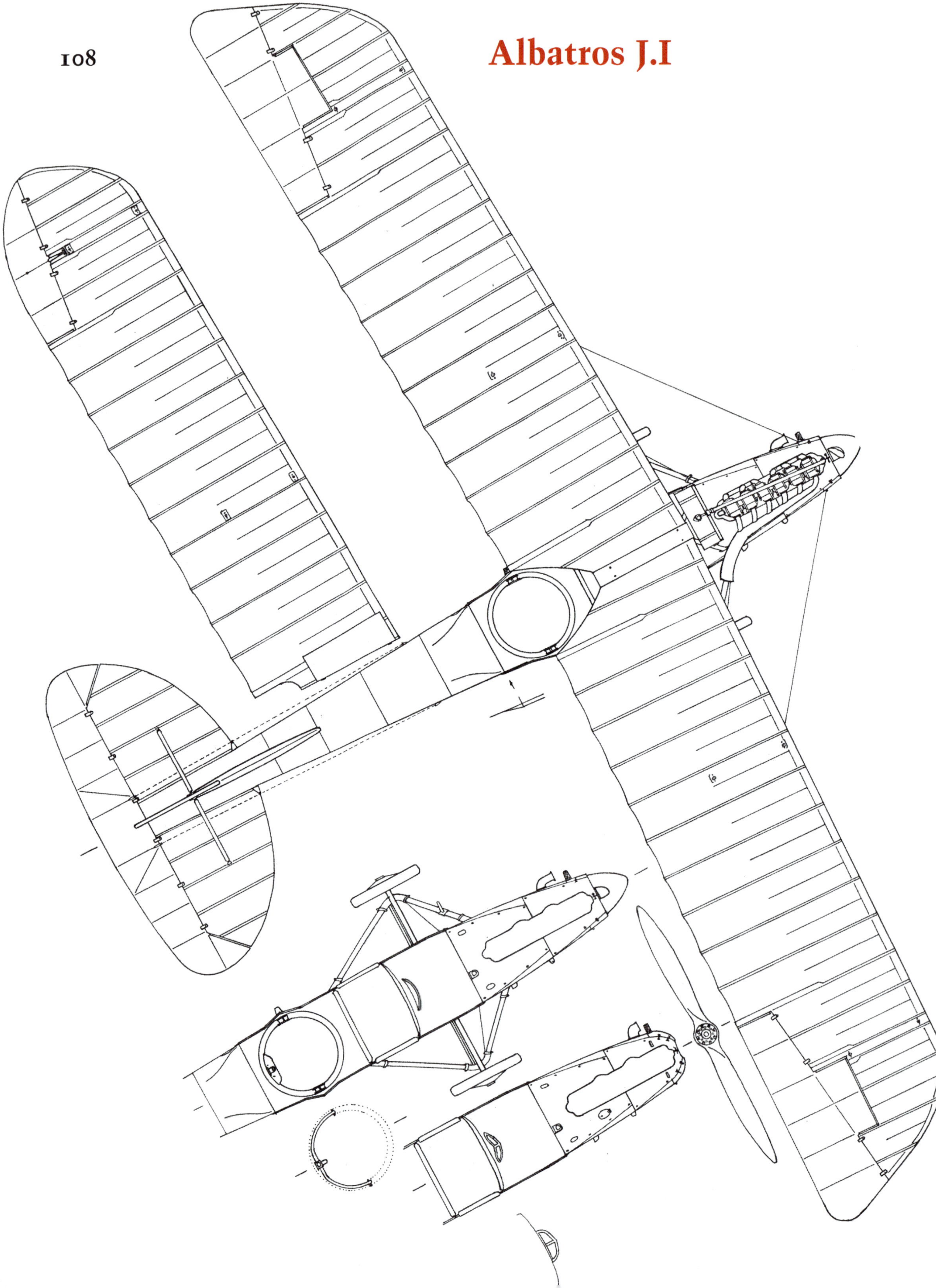

Albatros J.I

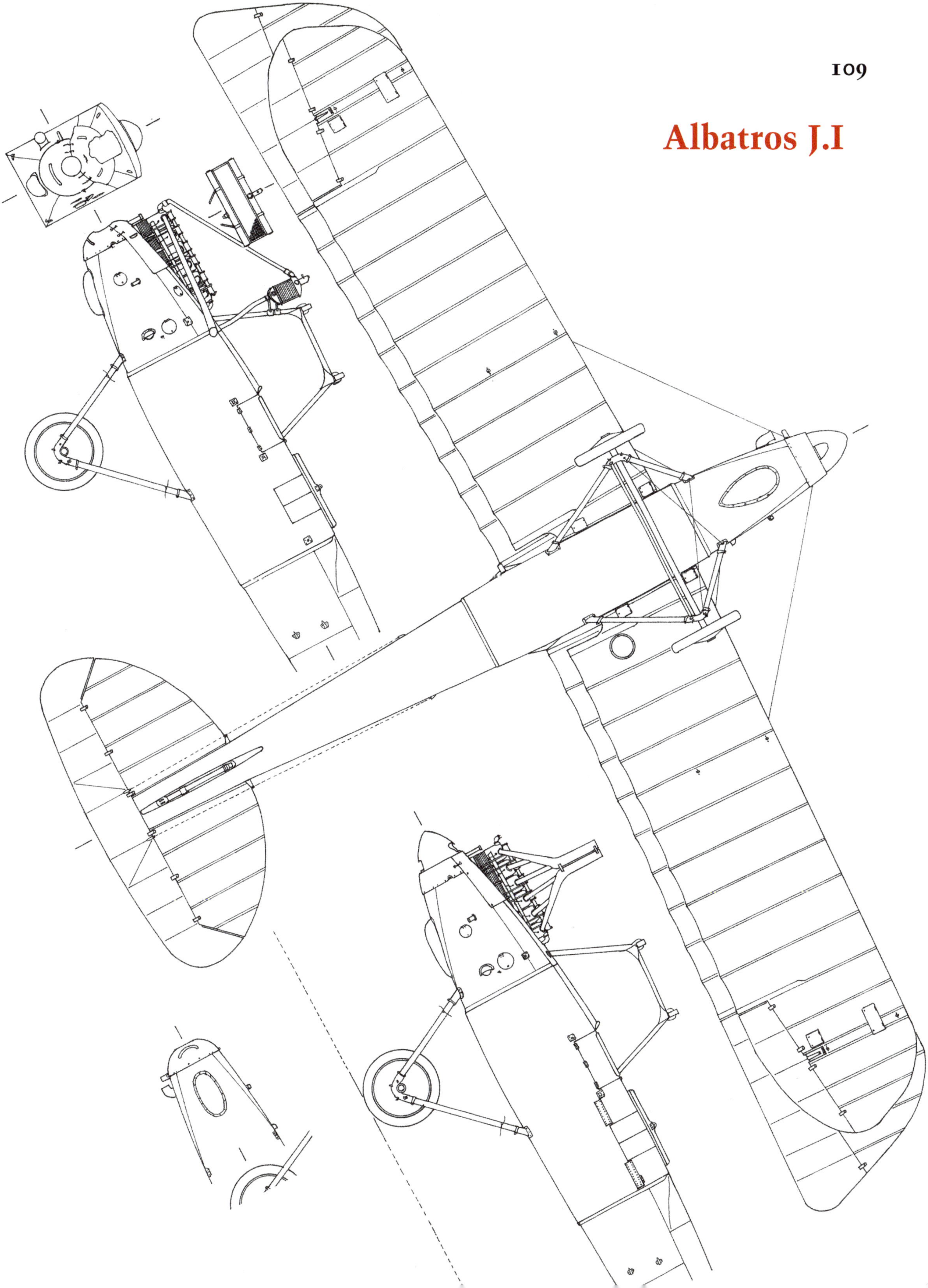

Albatros J.I

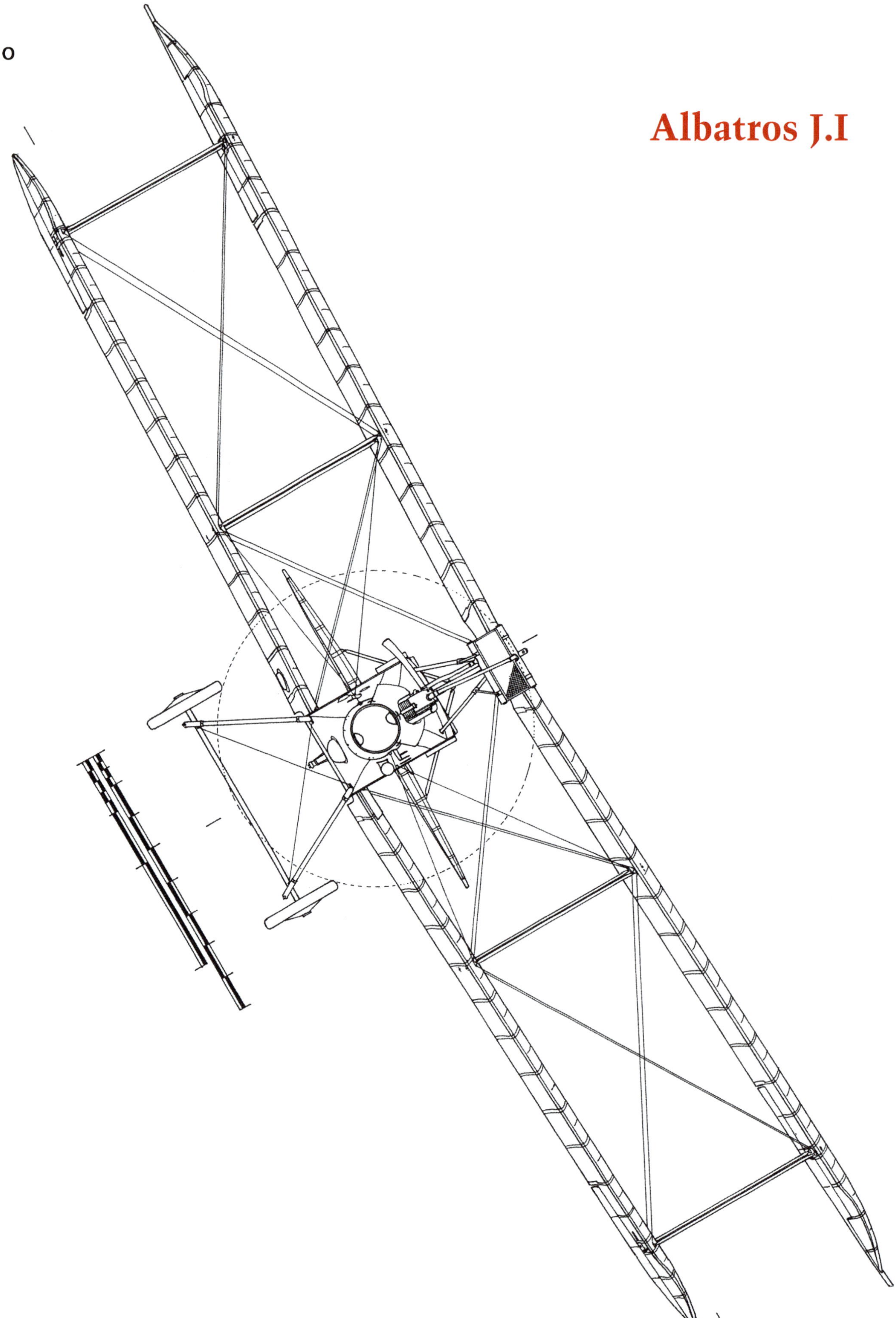

Albatros J.I

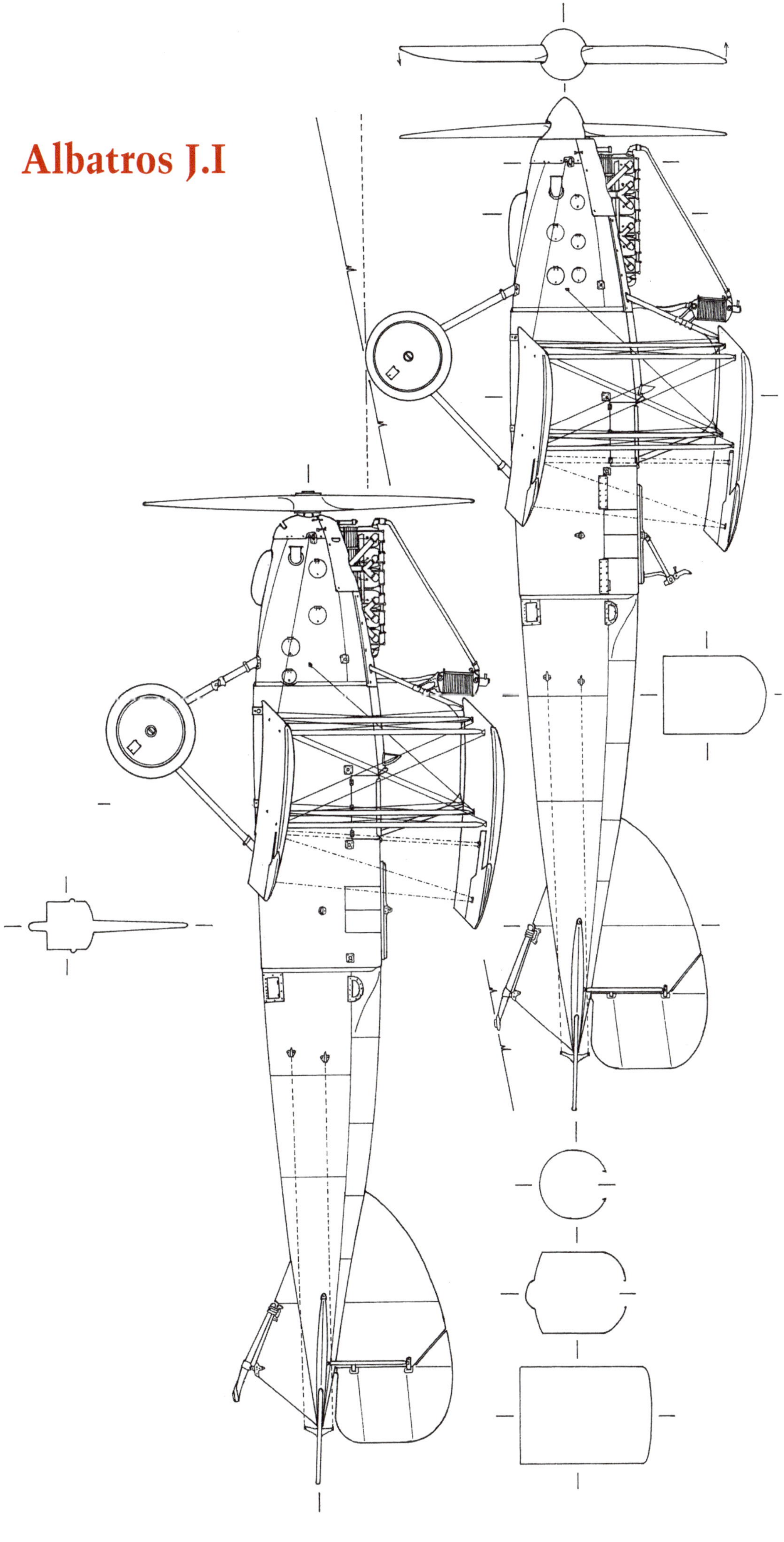

Albatros J.II

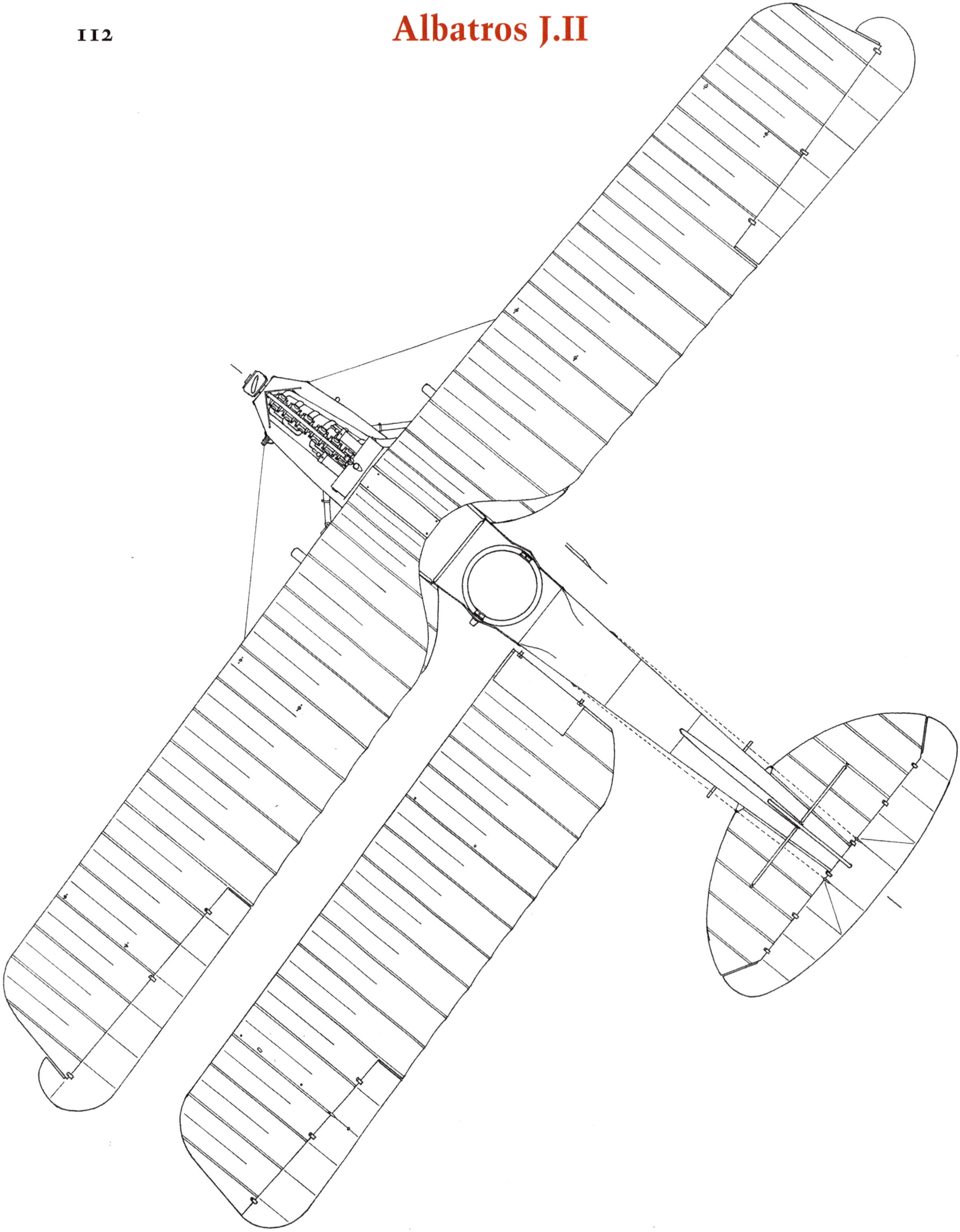

Albatros J.II

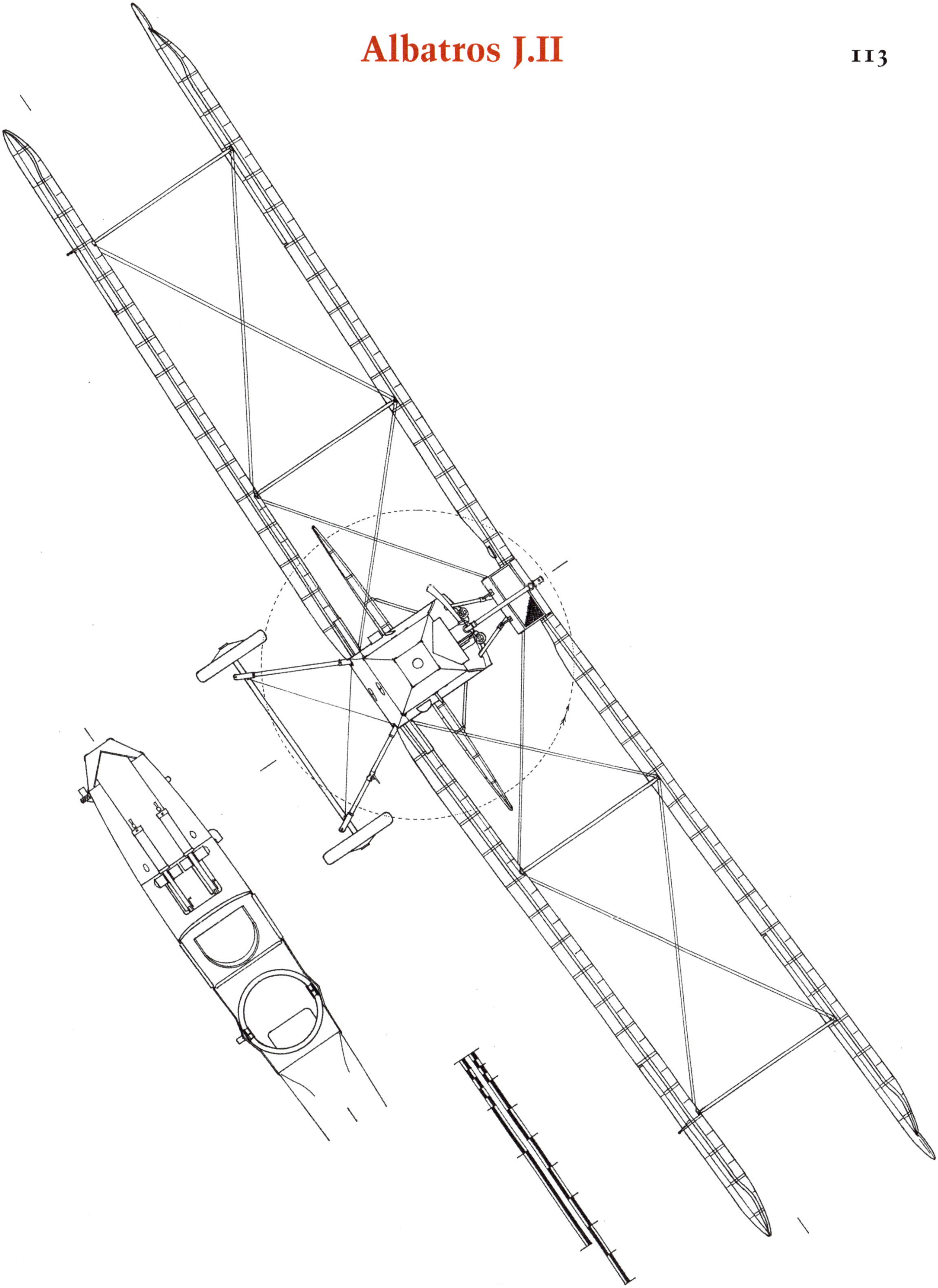

Albatros J.II

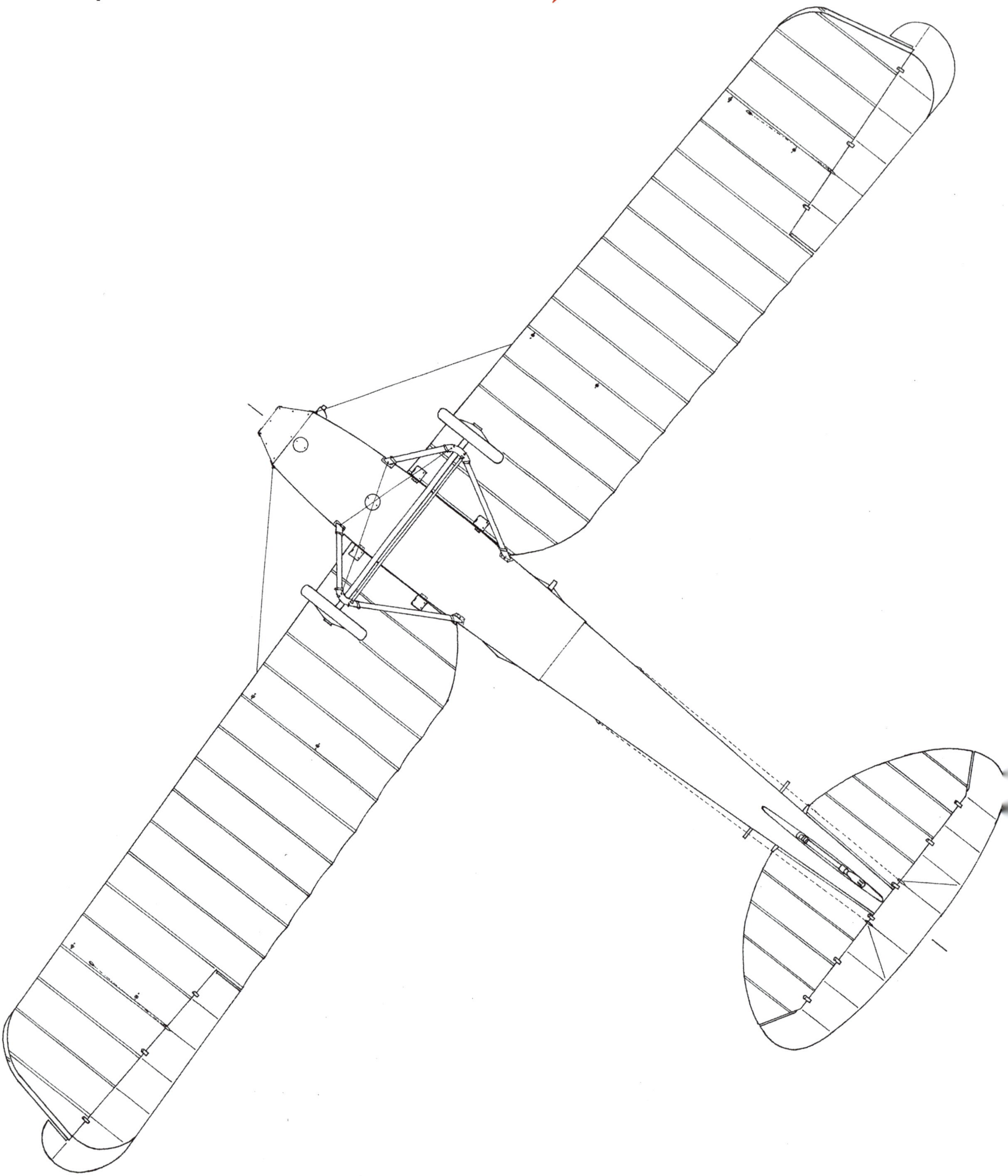

Albatros J.II

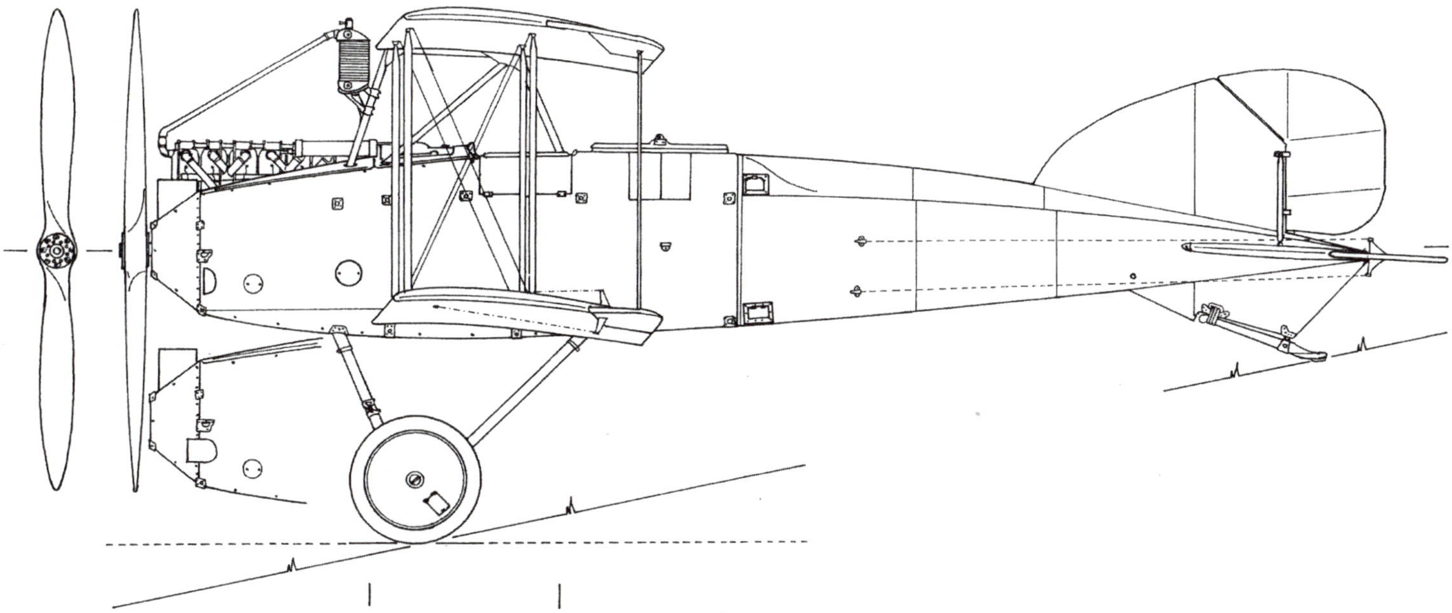

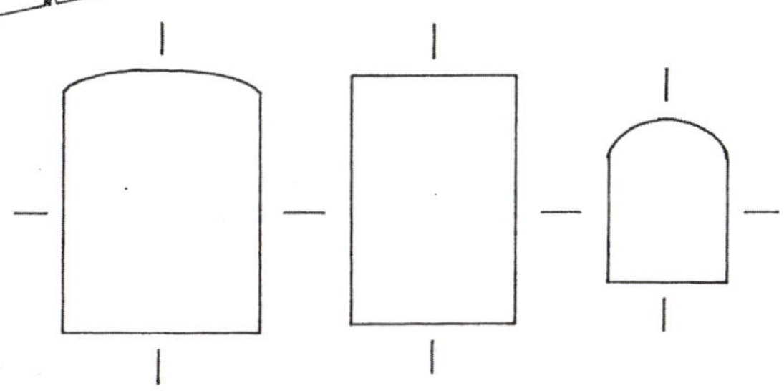

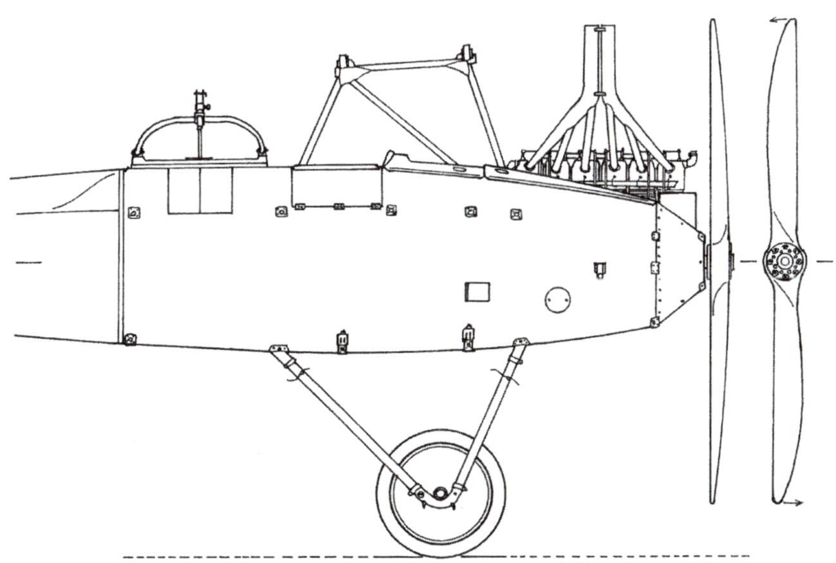

Afterword

Despite being used in modest numbers, the German armored warplanes were very effective in their designed infantry contact role and evolved into effective ground-attack and anti-tank weapons. There were no Allied counterparts to these unusual airplanes. Unfortunately, material shortages limited them to the 200 hp Benz Bz.IV instead of the more powerful engines they needed for better performance, maneuverability, and flight safety. Nevertheless, they pioneered specialized armored ground-attack aviation used so effectively in later conflicts.

Above: A group of Turkish officers tours an *Abteilung*; the excited aircrew are on the right and their AEG J.I 428/17 forms an interesting background.

Left: A truck retrieves what is left of a downed AEG J-type; given the sprayed camouflage it may be a J.I or early J.II. Although the rest of the aircraft is destroyed, the armored forward fuselage enclosing the engine and cockpits is still intact.

Below: Likely the wreck of AEG J.II 249/18 brought down in US lines on 27 October 1918.

Above: A downed AEG J-type; serial apparently J.2x6/xx. While resistant to ground fire, the AEG J-types were not as robust as the Junkers J-types.